Hatoon Baazim

CRISPR/dCas9:TF Mediated Transcriptional Regulation in Plants

Hatoon Baazim

CRISPR/dCas9:TF Mediated Transcriptional Regulation in Plants

LAP LAMBERT Academic Publishing

Impressum / Imprint

Bibliografische Information der Deutschen Nationalbibliothek: Die Deutsche Nationalbibliothek verzeichnet diese Publikation in der Deutschen Nationalbibliografie; detaillierte bibliografische Daten sind im Internet über http://dnb.d-nb.de abrufbar.

Alle in diesem Buch genannten Marken und Produktnamen unterliegen warenzeichen-, marken- oder patentrechtlichem Schutz bzw. sind Warenzeichen oder eingetragene Warenzeichen der jeweiligen Inhaber. Die Wiedergabe von Marken, Produktnamen, Gebrauchsnamen, Handelsnamen, Warenbezeichnungen u.s.w. in diesem Werk berechtigt auch ohne besondere Kennzeichnung nicht zu der Annahme, dass solche Namen im Sinne der Warenzeichen- und Markenschutzgesetzgebung als frei zu betrachten wären und daher von jedermann benutzt werden dürften.

Bibliographic information published by the Deutsche Nationalbibliothek: The Deutsche Nationalbibliothek lists this publication in the Deutsche Nationalbibliografie; detailed bibliographic data are available in the Internet at http://dnb.d-nb.de.

Any brand names and product names mentioned in this book are subject to trademark, brand or patent protection and are trademarks or registered trademarks of their respective holders. The use of brand names, product names, common names, trade names, product descriptions etc. even without a particular marking in this works is in no way to be construed to mean that such names may be regarded as unrestricted in respect of trademark and brand protection legislation and could thus be used by anyone.

Coverbild / Cover image: www.ingimage.com

Verlag / Publisher:
LAP LAMBERT Academic Publishing
ist ein Imprint der / is a trademark of
OmniScriptum GmbH & Co. KG
Heinrich-Böcking-Str. 6-8, 66121 Saarbrücken, Deutschland / Germany
Email: info@lap-publishing.com

Herstellung: siehe letzte Seite /
Printed at: see last page
ISBN: 978-3-659-61123-0

Zugl. / Approved by: Thuwal-Jeddah, King Abdullah University for Science and Technology (KAUST), 2014

ABSTRACT

CRISPR/dCas9:TF Mediated Transcriptional Regulation In Plants

Hatoon Baazim

Developing targeted genome regulation approaches holds much promise for accelerating trait discovery and development in agricultural biotechnology. Clustered Regularly Interspaced Palindromic Repeats (CRISPRs)/CRISPR associated (Cas) system provides bacteria and archaea with an adaptive molecular immunity mechanism against invading nucleic acids through phages and conjugative plasmids. The type II CRISPR/Cas system has been adapted for genome editing purposes across a variety of cell types and organisms. Recently, the catalytically inactive Cas9 (dCas9) protein combined with guide RNAs (gRNAs) were used as a DNA-targeting platform to modulate the expression patterns in bacterial, yeast and human cells. Here, we employed this DNA-targeting system for targeted transcriptional regulation *in planta* by developing chimeric dCas9-based activators and repressors. For example, we fused to the C-terminus of dCas9 with the activation domains of EDLL and TAL effectors, respectively, to generate transcriptional activators, and the SRDX repression domain to generate transcriptional repressor. Our data demonstrate that the dCas9:EDLL and dCas9:TAD activators, guided by gRNAs complementary to promoter elements, induce strong transcriptional activation on episomal targets in plant cells. Moreover, our data suggest that the dCas9:SRDX repressor and the dCas9:EDLL and dCas9:TAD activators are capable of markedly repressing or activating, respectively, the transcription of an endogenous genomic target. Our data indicate that the CRISPR/dCas9:TFs DNA targeting system can be used in plants as a functional genomic tool and for biotechnological applications.

.

TABLE of CONTENTS

LIST OF ABBREVIATIONS

GOI	Gene of interest
Cas9	CRISPR-associated endonuclease 9
CHS	Chalcone synthase gene
CRISPR	Cluster Regulatory Intersperced Palindramic Repeat
dCas9	dead-Cas9 (catalytically inactive Cas9)
DSB	double-stranded break
HD	Histidine, Aspartate
HDR	Homology directed repair
KRAB	Krüpple-associated box
NHEJ	Non-homologous end joining
NI	Asparagine, Isoleucine
NK	Asparagine, Lysine
NG	Asparagine, Glycine
NN	Asparagine, Asparagine
NS	Asparagine, Serine
PAM	Protospacer adjacent motif
RT-PCR	Reverse Transcription-Polymerase Chain Reaction
RVD	Repeat Variable Di-residue
TALE	Transcription Activator-like Effector
ZFN	Zinc Finger Nuclease

LIST OF ILLUSTRATIONS

LIST OF TABLES

1 Introduction:

The ability to perturb the transcriptional patterns and interrogate gene functions in a site-specific manner will open new bioengineering possibilities with applications in basic biology and biotechnology. Since the advent of plant biotechnology, researchers have attempted to up-regulate or down-regulate the expression levels of a specific gene of interest (GOI) for functional studies. Over-expression is one of the first techniques adopted in the investigation of gene function, enhancing the expression of the GOI and then examining the resulting phenotypic effects. The first applications of over-expression in plants used cell cultures [1] and later on *Arabidopsis thaliana*. These experiments included the activation-tagging system [2], followed by a study on the FOX-hunting system, which involved cDNA expression under strong promoter such as the cauliflower mosaic virus 35S promoter [3].

An earlier study during 1990 aimed to over-express Chalcone synthase in Petunia flowers, using a construct containing the petunia CHS gene expressed under the control of 35S promoter [4], unexpectedly resulted in the complete and/or partial blockage of the corresponding endogenous gene expression, producing a phenotype of white and/or patterned flowers. This study introduced a new technique, henceforth termed co-suppression, which was then widely used in Arabidopsis, Drosophila and C. elegance [5,6]. Studies investigating the co-suppression mechanism of action suggested two hypotheses, the first of which suggests that the silencing occurs transcriptionally (TGS), as a result of direct interaction between the chimeric gene and its endogenous homologue, which epigenetically silences the gene expression, either by methylation or chromatin modification. The second and more plausible hypothesis, suggested post-transcriptional gene silencing (PTGS), which allows the GOI to be transcribed, but prevents the gene transcripts from accumulating as a result of rapid RNA degradation machinery [7,8]. This process has a dosage effect; meaning that the RNA degradation is triggered when the number of transcripts of the gene exceeds a certain threshold, as a result of the expression of both introduced and endogenous homologue. Thus, the efficiency reduces when the endogenous homologue is not expressed or mutated to the extent of losing its functionality [9,10]. The RNA degradation here is very similar to RNA interference (RNAi) in animals. Both processes involve the formation of a dsRNA, sharing sequence homology with the co-suppressed gene, that are then processed to produce 20-25nt-long short-interfering RNA (siRNA) fragments [11]. These fragments then bind to the gene transcripts in a sequence-specific manner and subsequently mediate their degradation. Such conventional techniques, however, suffer from several limitations, such as their inefficiency, lack of reproducibility, and the random nature of their modification [12].

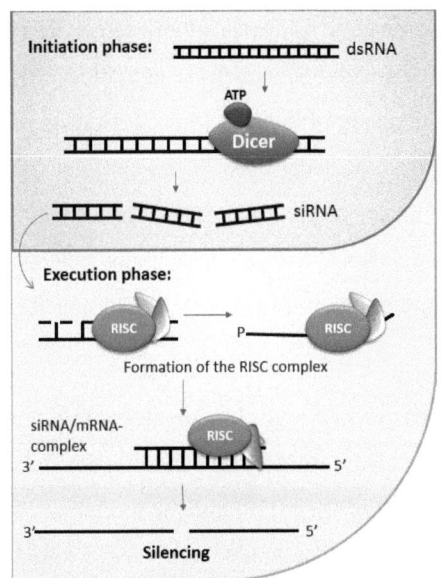

Figure 1: RNAi. Initiation phase: Multifunctional protein, Dicer, mediates the cleavage of dsRNA molecule carrying the sequence homology to the target site, this cleavage results in short RNA molecules called siRNAs. Excusion phase: siRNA molecules bind to the RISC complex, which results in the degradation of the siRNA sense strand. The remaining strand guides the RISC complex to mRNA targets and mediates their binding, subsequently leading to a rapid RNA degradation and gene silencing.

These co-suppression studies drew the attention to the RNAi system due to its sequence-specific binding properties. The first experiments introducing RNAi as a transcriptional modulation tool was performed in *C. elegans*, where they showed that antisense RNAs processed from a transgene or directly injected into a worm's gonad are capable of silencing genes in a sequence-specific manner [13]. Closer investigation of RNAi revealed that this process occurs in two phases, an initiation phase, followed by an execution phase. During the initiation phase the dsRNAs are generated, they then bind to a multifunctional enzyme called Dicer via its dsRNA-binding domain. This enzyme also contains an RNA helicase and a catalytic RNAse III which will cleave the dsRNA into short-interfering RNAs (siRNA) in an ATP-dependent manner. During the execution phase these siRNA duplexes bind to a protein complex called RNA-dependent silencing complex (RISC). This binding cause the degradation of the siRNA sense strand, thus, enabling the antisense strand to recognize the complementary sequence in the target gene mRNA [14]. The RISC performs both exonuclease and endonuclease activities [15], within the cell's cytoplasm and results in the silencing of the target gene [16] (Figure 1). It has been proposed that these siRNAs are repeatedly used in the cell. On the other hand the "Random degenerative PCR" theory suggest that the siRNA strands are used as primers that binds to the target mRNA and generate more dsRNA fragments by the help of RNA-dependent RNA polymerase (RdRP), amplifying the RNAi signal [17].

RNAi provided researchers with a fast, inexpensive and high-throughput module for transcriptional repression; however, the success of the RNAi gene silencing depends on the formation of small RNAs by endoribonuclease Dicer [18] and the formation of the RISC complex including Argonaute proteins [19]. The functionality of the RNAi system also depends on the presence of the components of the silencing machinery, and thus cannot be applied in every organism. Moreover, RNAi is limited by several other factors such as its elevated level of toxicity in certain cell types, off-target effects, non-heritability over many generations and difficulty of multiplexing. Consequently, there was an increasing interest in generating site specific synthetic transcriptional regulators that can be customized to regulate the expression levels of any user-selected gene(s). During the past few decades several platforms were generated fusing functional transcriptional domains to DNA-binding domains, such as Zinc finger nucleases (ZFN) and Transcriptional activator-like effectors (TALE), two techniques that revolutionized the field of targeted genome engineering.

1.1 Zinc finger-based transcriptional regulators:

Zinc finger arrays have been used as a programmable DNA binding module fused to different transcriptional functional domains. The first of these domains to be fused to the zinc finger was the FokI endonuclease [20], and was thereafter popularly used in many studies that optimized the system in different organisms [21-24]. A zinc finger array is made of a tandem repeat of 3 to 6 zinc finger domains linked by a highly conserved linker sequence [25]. The repeat is capable of recognizing theoretically any user-defined sequence in the DNA (Figure 2), providing a genome-engineering tool suitable for a variety of application, ranging from pure scientific studies to applications in crop development. Each zing finger domain is about 30 amino acids long, and is structured in the form of a ββα domain [25-27]. The structure of the zinc finger domain is stabilized by a zinc-ion, from which it takes its name, coordinated between a Cysteine and Histidine residues [25]. When the zinc finger array comes in contact with the DNA, the α helix domain protrudes into the minor groove of the DNA double helix, and the amino acids on the N-terminus of the helix domain

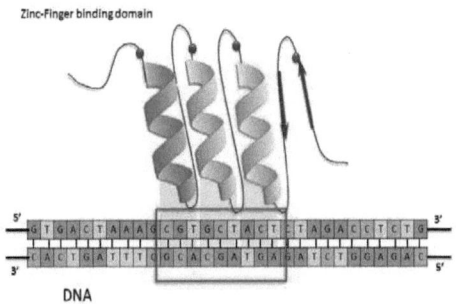

Figure 2: Each zinc finger is made of a ββα domain. Zinc finger arrays are assembled from 3 to 6 finger domains linked together using a linker sequence stabilized by a zinc ion. The N-terminus of each finger recognizes a 3bp code on the DNA strand, facilitating target recognition.

recognize 3 base pairs on the DNA strand [28] (Figure 2).

Natural transcription factors usually consist of at least two domains, a DNA-binding domain, and an effector domain, which performs the function of either activation or repression [29]. Fusing the effector domain to a different DNA binding domain has been proven to be an effective method in generating a versatile and flexible system to alter gene expression profiles. One of the first transcriptional activators fused to zinc finger array was the herpes simplex virus VP16 domain and its derivatives. VP16 is a strong activator that functions both in mammalian and plant systems [30]. It consists of 490 amino acids, and contains manly two regions, a transcriptional activation domain, rich with acidic residues, and a core domain, which directs the VP16-induced complex formation. The core domain of VP16 is highly conserved, in contrast to the activation domain that differ between different strains of the virus even though it interacts with many major components of the transcriptional machinery [31]. A fusion of four tandem repeats of the minimal VP16 activation domain with a 3-finger and 6-finger zinc finger domains was performed, using a luciferase (LUC) reporter gene. This study revealed high levels of activation induced by the VP16-ZF array, and tested several binding targets, shedding the light on the importance of the binding location of the transcriptional regulator [32].VP16-ZF complexes that targeted regions far from the TATA-box showed significant reduction in the level of activation. Other complexes targeting the beginning of the transcription start site conveyed no activity, conceivably because it blocked the binding of the RNA polymerase complex [32]. Designing such constructs with DNA binding domains directed at transcription initiation sites is one strategy by which transcriptional repression can be achieved, however, this kind of repression approach proved not to be very effective [27,33]. Transcriptional repressors can also be achieved by fusing repressor domains, such as Krüpple-associated box (KRAB), found naturally at the N-terminus of zinc finger proteins [34], or mSin3 interaction domain, which mediates repression by recruiting a histone deacetylase [35]. An important factor that plays a role in the efficiency of the activation or repression is the strength of the promoter of the target gene. It appears that strong promoters require strong repressors to repress their activity, and vice versa [32].

Transcriptional factors are not the only functional domains that can be fused to the zinc finger arrays. Zinc finger proteins have been fused to many other catalytic domains, such as recombinases [36,37], methylases [38,39] and transposases for targeted gene integration [40]. Zinc finger recombinases (ZFRs) contain a unique DNA binding site that recognize two inverted zinc finger binding sites, flanking a central sequence of 20 base pairs [41]. They are capable of mediating DNA cleavage, strand exchange and re-ligation without requiring DNA synthesis or high energy cofactors [42]. In addition to their autonomous functionality; unlike other nucleases, they are capable of cleaving their target and re-ligating independently of the cellular repair mechanisms, which enables them to work in nearly any cell during any developmental stage. However, these recombinases are laborious to construct and require significant re-engineering to achieve user-defined DNA targeting. Cytosine-5 DNA methyl-transferases (C5-Mtase) are enzymes that function by binding to a specific sequence in the DNA, it targets a specific cytosine molecule, flips it out into its nucleotide-binding pocket and transfers a methyl group from the cofactor S-adenosyl-L-methionine (AdoMet) to the position 5 of the cytosine [42]. Fused with zinc finger arrays C5-Mtase can be used in many applications. The methylation of cytosine in the CpG islands in the promoter region of a specific gene is capable of silencing the expression of that gene [43,44]. Moreover, the C5-Mtase ZF complex can be modified into a cytotoxic enzyme that binds specific locations in the DNA with extremely high affinity [45,46]. This modification is achieved through the mutation of a cysteine amino acid in the catalytic active site of the C5-Mtase into a glycine. This cytotoxic construct can have potential applications in targeted cell death.

Even though zinc finger DNA binding module provided a highly specific and versatile tool for targeted genome modification, the construction of such modules remained to be a formidable task. Different approaches have been developed to facilitate the engineering of zinc fingers as DNA binding module, but the process remains limited in terms of target diversity, and efficiency. The construction of these modules is also very laborious, time consuming and suffers from low reproducibility [47,48]. It is also important to further investigate potential off-targeting effects that could lead to cellular toxicity [20,49,50].

1.2 TAL-based transcriptional regulators:

Transcription activator-like effectors have been identified in phyto-pathogenic bacteria including *Xanthomonas* and *Ralstonia* where they reprogram the transcriptional machinery of the plant host to their own benefit [51]. TALEs are transcription factors and once injected via the type III secretion system (T3SS) they localize to the plant cell nucleus, where they bind to the regulatory elements of the target genes. Several TAL-based transcriptional regulators have been generated in a variety of organisms. These proteins are typically composed of three domains; the N-terminus translocation domain, the central repeat domain, and the C-terminus acidic transcriptional activation domain (Figure 3). Most TAL effectors contain central repeat domain composing between 13 and 28 repeats [52], each of which have nearly identical 33-35 amino acids, the polymorphism between these amino acids occurs in the position 12 and 13 called the repeat-variable di-residues (RVD). Each RVD mediates site-specific binding to one nucleotide in the DNA target sequence (figure 3). When using a truncated repeat region, artificial TALEs maintained their activity down to a minimal of 6.5 repeats, and the 10.5 repeats were sufficient for full induction of the target genes [53]. Curiously, all naturally occurring TALEs contain a truncated repeat of 20 nucleotides targeting a thymidine nucleotide. This repeat unit is called 0-repeat or half repeat. The conservation of this repeat could be an indicator

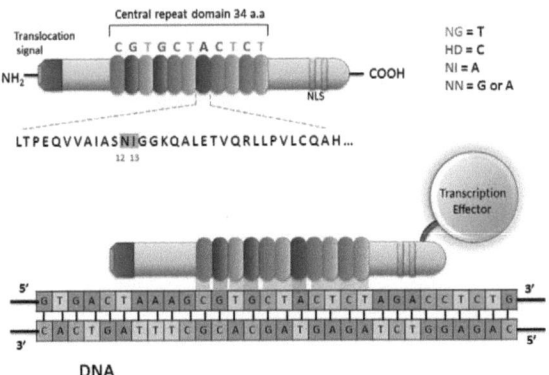

Figure 3: TAL effector DNA binding domain. In TALE binding, each repeat contains the 12 and 13 amino acids (RVDs) which recognize the DNA sequence in a RVD-nucleotide manner. TALE can be fused to either, transcriptional activators or repressor to regulate gene expression.

of either the evolutionary origin of TALE proteins, or of a critical functional necessity, possibly to maintain the integrity of the protein's structure, however, no evidence of this has been provided as of yet [54]. A recent study claimed that this half repeat is in fact dispensable, by designing TALE proteins lacking the 0-repeat domain for transient expression assays in *N. Banthamiana* and specific targeting of endogenous genes in rice. The results represented in the this study showed comparable levels of gene activation with or without the presence of the last half-repeat domain [54].

The code of repeat-to-nucleotide binding allows the facile engineering of any user-selected DNA sequence. The studies that investigated this code

RVD	Nucleotide
HD	C
NI	A
NK	G
NG	T
NN	A or G
NS	A,C,G or T

Table 1: RVD:nucleotide binding specificties.

used computational approaches to decipher the RVD-nucleotide code and identified several RVD specificities [51,55] (Figure 4, Table 1). RVDs retain different strengths in terms of affinity, some were classified as weak RVDs such as NI, NK and NG, which allow more flexible binding specificities, others such as HD and NN have stronger specificities [56], thus, designs with varying flexibility can be made, allowing for instance binding in different systems without the need for codon optimization.

Co-crystal structures of TALE DNA-binding domains bound to their target nucleotides revealed that individual repeats comprise a two-helix v-shaped bundles that stack to form super-helix around the DNA. The position of the hyper variable residues at position 12 and 13 are positioned in the DNA major groove, with the residue at the position 8 and position 12 within the same repeat interact with each other, thereby possibly stabilizing the structure of the domain while the residue at position 13 makes the base-specific contact with the DNA [57,58]. Since TAL effectors are natural transcription activators, studies implementing them for their activation function mainly tested the modularity of their DNA binding domain. A study using the scaffold of the TALE AvrBs3 successfully altered the specificity of this TAL effector by changing the RVD to match the tomato Bs4, and the *Arabidopsis* EGL3 and KNAT1 promoter [59]. Fusion of repressor domains such as SRDX to TAL effectors has been shown to repress the expression of the RD29::LUC transgene and the RD29A endogenous gene as well as several other genes in *Arabidopsis* [60].

As robust and efficient TAL-based synthetic transcriptional regulators are, they are still limited by several factors, such as the requirement of TAL protein engineering for every single target, the unpredicted off-target binding due to repeat-context effects and the need to express large proteins for every experimental set-up. Moreover, TALE repeats contain a large number of highly conserved sequences, which requires more advanced cloning methods, and the highly repetitive nature of TALEs creates restrictions in the carrier vectors used for their delivery [61].

1.3 CRISPR/Cas9 systems:

The CRISPR/Cas9 system (Cluster Regulatory Interspersed Palindromic Repeat) is a natural system used by about 40% of bacteria and 90% of archaea as a form of adaptive immunity against invading viral or plasmid DNA [62]. The natural bacterial system consists of Cas protein operons, CRISPR array, and two non-coding RNAs (Figure 5). The main functional domain in this system is the Cas protein, providing the nuclease activity. The CRISPR array on the other hand provides the DNA-binding specificity. It contains a number of conserved repeat domains that separates the spacer sequence, a sequence of 20 nucleotides complimentary to segments of the foreign viral or plasmid DNA (Protospacer) [63,64]. Upon initial exposure to the viral or plasmid invader, the surviving bacteria copy segments of the protospacers and incorporate them into its CRISPR array. This sequence is utilized during a secondary exposure to the same viral or plasmid genetic material. In type II CRISPR/Cas system, the CRISPR array is transcribed into premature CRISPR RNA (pre-crRNA), a long array containing multiple spacer sequences. This pre-crRNA is then processed, by the trans-acting CRISPR RNA (tracrRNA) and RNAse III, into a fully matured short crRNA [65-67], providing a guide to Cas9 endonuclease in a simple Watson-and-Crick sequence complimentary manner, resulting in the cleavage of the foreign DNA by the Cas9 endonuclease (Figure 5). Cas9 endonuclease differentiates between self and non-self DNA through a tri-nucleotide sequence (NGG), referred to as protospacer-adjacent motif (PAM), a motif present only on the foreign DNA and is recognized by the PAM-interacting domain (PI domain) in the Cas9 protein [68]. Unlike type I and III CRISPR/Cas system, where the nuclease activity is performed by a multi-Cas protein complex, type II CRISPR/Cas requires only Cas9 endonuclease.

The modularity of the DNA-binding domain of this system, and the simplicity of its construction provided a fast and efficient tool for targeted genome engineering across prokaryotic and eukaryotic species [69,70]. By fusing the two RNA molecules into a chimeric single guide RNA (gRNA), containing a mature crRNA and a partial tracrRNA the system have been developed to require only two components, a Cas9 protein and a chimeric single-guide RNA (gRNA). In principle, using this system, it is possible to target any user-defined sequence in the DNA and introduce a double stranded break (DSB), which would trigger the cellular repair machineries such as non-homologous end joining (NHEJ) and homology directed repair (HDR), with the only one requirement being the PAM sequence.

A recent study investigated the crystalline structure of the CRISPR/Cas9 system during its binding to the target DNA in *S. pyogenes* with a resolution of 2.5A [71]. This structural analysis revealed that the Cas9 protein consists of two lobes, a recognition lobe (REC), and a nuclease lobe (NUC), with the gRNA:DNA heteroduplex contained in a positively charged groove in between the two lobes. This

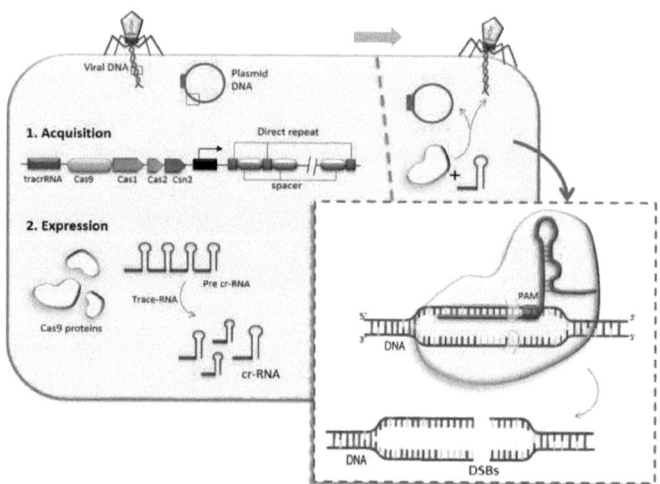

Figure 4: 1. Acquisition phase: When a foreign viral or plasmid DNA invades the bacterial cell, surviving bacteria acquire fragments of the foreign DNA (protospacer) and inserts it into its CRISPR array as spacer sequences separated by conserved repeat sequenc. 2. Expression: upon secondary exposure the spacer sequences are transcribed into a pre-crRNA, alongside the tracrRNA and Cas9 protein. The tracrRNA mediates the maturation and cleavage of the pre-crRNA into crRNAs which is them associated with the Cas9 protein to bind to the invading viral or plasmid DNA and introduces DSBs into the foreign DNA.

study also confirms the critical role of PAM sequence in the recognition and binding to the DNA target [70,72], and identified the PAM interacting domain (PI), which recognizes the NGG sequence. This study also showed that the PI domain facilitates the separation of the DNA strands and the formation of the gRNA:DNA heteroduplex.

These targeted genome modulation technologies, such as ZFs, TALEs and CRISPR/Cas systems, provide a new generation of tools to answer core biological questions that could include the DNA repair mechanism, recombination, metabolism and stress response. These systems could also have medical application, through their capability of introducing correct mutations of genetic diseases and supplying the correct template for DNA repair pathways to adopt and re-write the mutated sequence. Using these targeted genome modulation techniques has the potential to remove concerns about the insertion of foreign DNA into natural systems and the random DNA integration process. This would help in casting away the doubt of GM crops in the eyes of the public. These new biotechnological advances would also improve the overall quality and quantity of current crop production, accelerate beneficial trait development and expand the range of traits.

1.3.1 RNA-guided dCas9 transcriptional regulation:

The NUC lobe of the Cas9 contains the Cas9 catalytic domains, RuvC and HNH [71]. Each domain catalyzes the cleavage of a single DNA strand; RuvC domain cleaves the sense strand, and the HNH domain cleaving the anti-sense strand. Mutations introduced to either catalytic domain converts the Cas9 protein into a nickase [70,73], which

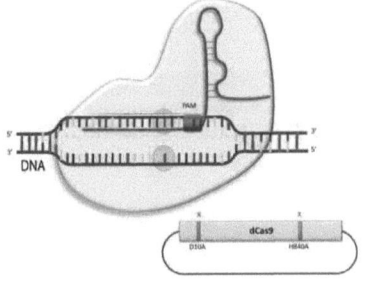

Figure 5: A schematic representation of the dCas9 protein, point mutations in the Cas9 catalytic sites (D10A and H840A) results in a catalytically inactive Cas9, referred to as dead Cas9 (dCas9) protein.

promotes homology-directed repair (HDR). This is particularly useful due to the higher fidelity repair of this mechanism, and the difficulty of promoting HDR in flowering plants where NHEJ is the dominant process [74]. Introducing mutations in both catalytic domains (D10A, H840A) produces a catalytically inactive Cas9, or dead Cas9 (dCas9) incapable of modifying the DNA (Figure 6). However, this dCas9 together with a sgRNA is capable of generating a DNA-binding complex that recognizes a sequence in the DNA. Therefore, catalytically inactive Cas9 (dCas9) can be used as a DNA targeting module using user-specific crRNA sequence.

1.3.2 Transcriptional regulation in bacteria:

CRISPR/dCas9 systems have been applied to several bacterial organisms, such as *E.coli, Streptococcus pyogenes* and *Streptococcus pneumonia*. In *E.coli*, targeting a dCas9 protein to the promoter elements or coding regions of the target genes was sufficient to significantly repress the gene expression in a site-specific manner [75]. The transcriptional repression effects could be due to a steric blockage of the binding of RNA Pol II or inhibition of the transcript elongation. The efficiency of the transcriptional repression is affected by the gRNA binding distance in the promoter from the transcription start site of the target gene. Closer binding from the transcriptional start site produces significant transcriptional repression. It remains to be determined whether the distance from the transcriptional start site is promoter dependent or not. Other factors affecting the efficiency include the complexity of the genome and its epigenetic status, which affects the accessibility of the DNA to the CRISPR/dCas9 system.

1.3.3 Transcriptional regulation in eukaryotes:

CRISPR/Cas transcriptional regulators have been tested in several eukaryotic systems, including yeast [76], tobacco [77,78], wheat [79], mice [80,81], rat [82], rabbits [83], frogs [84], and fruit flies [85,86] as well as others. In yeast, CRISPR/Cas systems were used to increase HDR and showed 100% rates of recombination of donor DNA sequences [76]. The study in yeast also showed potential mutagenic capabilities of CRISPR/Cas9 systems by targeting an endogenous gene CAN1, a gene encoding a membrane arginine permease. This gene was chosen because it provides a negative selectable marker on its own, without the need to integrate a reporter gene to the system. When yeast cells are grown on a media containing the toxic arginine analogue canavanine, after transfection of the Cas9 endonuclease and the gRNA targets [87], surviving cells will have (scar)-sequences, left after cellular repair of the DSBs generated by the CRISPR/Cas9 complex, preventing the permease gene from being expressed.

A dCas9 system has been developed recently to modulate gene expression in yeast and mammalian cells. Mammalian cells however, have a very intricate genome that is regulated by a variety of epigenetic modifications and regulatory pathways, and the dCas9 alone showed very little efficiency [75]. Thus, attempts have been made to improve the transcriptional modulation by fusing functional domains that might recruit transcriptional activators or repressors. For transcriptional repression, KRAB and Kox1 were fused to dCas9 and successfully achieved gene repression. Furthermore, transcriptional activators such as VP16 and p65 activation domain were also fused to dCas9 and modulated transcriptional activation in both yeast and human HEK293 cells [88].

A recent study used mouse and human HEK293T cells to establish a CRISPR activator system [89], by fusing a minimal VP16 transcriptional activator domain (VP48) to the C-terminus of dCas9 protein. In their transient expression assay they used the GFP transgene reporter system in mouse embryos and a tdTomato reporter transgene in human HeLa cells to show dCas9VP48 activation of the promoter sequence of these genes. They also demonstrated the synergy and modularity of the activation level through by transfecting HEK293T cells with the dCas9 activator and serial of gRNA dilutions. The activation of the human endogenous gene IL1RN promoter in HEK293T cells however required a stronger activator; the VP16 was used instead of VP48, targeting multiple sites on the IL1RN promoter. Notably, they reported that targeting a dCas9 activator downstream of the transcriptional start site hindered the gene expression, presumably by blocking the polymerase enzyme and thus interrupting elongation.

1.3.4 Development of synthetic transcriptional regulators in plants via dCas9 CRISPR systems:

Here, we attempted to optimize and use the dCas9 as a platform for targeted genome regulation in plant cells. We used the dCas9 as a DNA-targeting module that can be reprogrammed by the engineering of the gRNAs. Although the dCas9 per se might function as a repressor, we sought to generate different chimeric proteins with functional domains that have been proven to mediate transcriptional activation or repression. For example, to generate a synthetic transcriptional activator we fused the dCas9 backbone to the activation domain of the dHax3 TAL effector protein to generate the dCas9.TAD hybrid protein. Similarly, we generated another synthetic activator by fusing the EDLL domain to generate the dCas9.EDLL. These chimeric dCas9 transcriptional regulators would allow us to test whether this system can be applied for site-specific transcriptional activation in planta. To generate synthetic repressors, we fused the SRDX domain to the dCas9 backbone to generate the dcas9.SRDX protein.

Both the EDLL and SRDX domains are found in plant transcription factors. They are both associated with the ERF/EREBP family of transcriptional regulators. The ERF/EREBP family members regulate transcription by binding to the GCC box on the promoter of their target gene, and they are involved in the regulation of important plant processes, such as tolerance of biotic and abiotic stress and development [90-

92]. EDLL is a relatively short motif, consisting of 24 amino acids. It is an acidic activator, comprising several acidic residues dispersed amongst hydrophobic leucine residues [93]. EDLL activates gene expression by binding to either proximal or distal positions from the TATA-box of their target gene, with an activation strength that varies with the position of their binding. SRDX on the other hand is driven from the transcriptional repressor domain EAR (ERF-associated amphiphilic repression domain) [94]. This domain has been found to be a very strong repressor, capable of performing its activity even in the presence of strong activators such as VP16 [95]. To test the functionality of our synthetic transcriptional regulators, we used the Agro-infiltration transient assay system to deliver effector and target molecules in tobacco leaves (figure 7). Our results show that the synthetic activators and repressors exhibit strong gene regulation activities. Our suite of synthetic transcriptional regulators may have broad applications in plant biotechnology and may help identify novel agricultural traits in important crop species.

2 Materials and methods:

2.1 Plasmid construction:

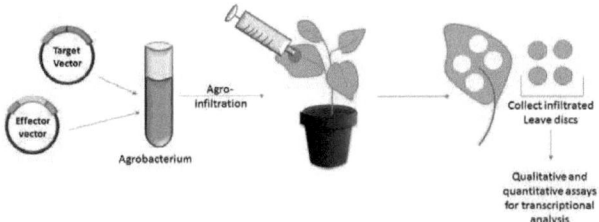

Figure 6: Overview of the experimental design. Agrobacterium tumefaciens GV3101 strains harboring gRNA target sequences (U6::gRNA) and dCas9 chimeric effector constructs (35S::dCas9, 35S::dCas9:.TAD, 35S::dCas9:.EDLL and 35S::dCas9:.SRDX) were mixed, in different combinations, and co-delivered into 3 to 4 weeks old Nicotiana benthamiana leaves. 36 to 48 hours post infiltration, leaf discs were collected and subjected to qualitative and quantitative analysis. When testing the activity of the transcriptional regulators on an episomal target Agrobacterium GV3101 strains harboring the Bs3::uidA in pKGWFS7 is mixed with effector and gRNA molecules.

Human codon optimized dCas9 was obtained from AddGene (44246). dCas9 was amplified from the original plasmid with primers 5'dCas9-F and 3'dCas9-R (SI) and subcloned into pENTR-D/TOPO vector (Invitrogen). dCas9 clone was verified by Sanger sequencing using primers listed in Supplementary information. dCas9 was subcloned to the destination vector pK2GW7 by the gateway LR reaction using LR clonase II (Invitrogen). Expression of dCas9 was driven by the constitutive cauliflower mosaic virus promoter 35S.

To generate the dCas9 fusions with transcriptional activators and repressors, fragments for dCas9 C-terminus in frame fusions of EDLL, TAD and SRDX were custom synthesized by BluHeron (SI).

To generate dCas9.EDLL activator we custom synthesized a fragment comprising the C-terminus end of dCas9 fused to EDLL domain flanked by MluI and EcoRI restriction sites (SI). The EDLL restricted fragment was subcloned to dCas9/pK2GW7 restricted with MluI and EcoRI to generate the intended C-terminus fusion. A similar strategy was used to generate dCas9.SRDX fusion protein using MluI and EcoRI (SI). Due to the presence of EcoRI restriction site within the TAD sequence we custom synthesized a fragment comprising a dCas9 C-terminus fused to the TAD flanked by MluI and XhoI restriction sites (SI). MluI and XhoI were used to directly exchange the EDLL domain sequence in the dCas9/pK2GW7 with TAD.

A chimeric RNA, comprising crRNA and tracrRNA, was used in this study. Chimeric RNA backbones were custom synthesized and expressed under the control of U6 Arabidopsis thaliana RNA polymerase III promoter (SI). All chimeric RNA backbones comprising U6 promoter, 20nt targets and gRNA sequence were custom synthesized (SI). The chimeric RNA containing specific targets was PCR amplified with the forward primer carrying BamHI restriction site and reverse primer containing recognition sequence for XbaI (SI). The integrity of the PCR product was confirmed by Sanger sequencing using primers listed in the supplementary material.

These fragments were then cloned into the multiple cloning site (MCS) in the pYL156 destination vector by restriction-ligation reaction using BamHI and XbaI. mBs3 promoter was fused to uidA gene in pKGWFS7 vector as described before.

2.2 Agroinfiltration:

All dCas9 chimeric transcriptional regulators subcloned into pK2GW7, U6.gRNA cloned in pYL156 and the Bs3.uidA in pKGWFS7, were transformed into A. tumefaciens, strain GV3101 by electroporation [REF]. Agrobacteria were first cultured overnight in 5 mL LB media containing suitable antibiotics, and subsequently cultured in 20 mL LB and grown to the OD600 between 1.0 and 1.5. The Agrobacteria were collected and re-suspended in the infiltration buffer (10mM MgCl2, 5mM MES, 0.1mM acetosyringone). To study effector target binding and activation or repression, combination of Agrobacteia containing effectors, episomal targets and gRNAs were mixed together to the OD600 0.6 and co-delivered in different combinations into three to four week old tobacco N. benthamiana leaves.

2.3 GUS qualitative and qualitative assays.

Infiltrated leaves were collected 36 to 48h post-infiltration for qualitative assay, and immersed into GUS staining buffer (10mM NaH_2PO_4, 10mM EDTA, 0.1% Triton-X, 0.1% X-gluc, 1mM $K_4Fe(CN)_6$, 1mM $K_3Fe(CN)_6$) and kept at 37°C for 24h. The following day discs were de-stained with 70% ethanol. For quantitative assays, proteins were extracted from the disc leaves with 150µl of GUS extraction buffer (50mM NaH_2PO_4, 10mM EDTA, 10mM β-mercaptoethanol, 0.1% Triton-X, 0.1% SDS). The protein amounts in the samples were quantified by Bradford assay. For the fluorometric assay, 90µl of assay buffer (50mM NaH_2PO_4, 10mM EDTA, 10mM β-mercaptoethanol, 0.1% Triton-X, 0.1% SDS, 5mM MUG) was mixed with 10µl of sample and incubated at 37°C for 60 min. Reaction was stopped by adding 900µl 0.2M sodium carbonate (pH 9.5). Fluorescence was measured in a Tecan microplate reader at 360 nm (excitation) and 465 nm (emission) with 4-methyl-umbelliferon (MU) dilutions as standard. In all experiments, background fluorescence, or auto-fluorescence, was subtracted from the fluorescence readings of the samples.

2.4 RNA isolation and semi-quantitative PCR.

Total RNA was isolated from the leaf discs using the Qiagen RNeasy Plant mini kit. Approximately 2 µg of total RNA was used for reverse transcription with SuperScript First-Strand Synthesis System for RT-PCR (Invitrogen), to generate cDNA. ActinI and N.b PDS gene was PCR amplified using Phusion High-Fidelity DNA Polymerase (Promega) with primer sets NB.actinI_RT_F, NB.actinI_qRT-Liu_R and qRT.NB.PDS.303F, qRT.NB.PDS.875R, respectively (SI). DNA concentration of samples was adjusted to equal molar ratios using actin as a reference.

3 Results:

3.1 Design and construction of dCas9-based chimeric transcriptional regulators.

To test whether dCas9 can be used as a DNA targeting module for site-specific transcriptional modulation in plants, we designed and constructed chimeric dCas9-based transcriptional activators and repressors, respectively. The human codon-optimized dCas9 was PCR amplified from pdCas9-humanized plasmid [75], subcloned into pENTR-D/TOPO and subsequently cloned into pK2GW7 plant expression vector by LR Gateway recombination cloning. The pK2GW7/dCas9 was used as a backbone to generate dCas9 C-terminus fusions with functional transcriptional activation and repression domains respectively. To build a chimeric transcriptional activator, we selected the EDLL domain from the ERF/EREBP family of transcriptional regulators because it was shown to function as a strong activation domain and be transferable to other proteins and active in proximal and distal positions from the target promoter [93]. To generate the dCas9.EDLL chimeric transcriptional activator, we fused a custom synthesized DNA fragment encoding the EDLL domain (EVFEFEYLDDKVLEELLDSEERKR) in frame to the C-terminus of dCas9 backbone using MluI and EcoRI restriction enzymes (Figure 7). Because TAL effectors functions as transcription factors in planta, we constructed a second transcriptional activator by fusing the activation domain of the dHax3 TAL effector to the C terminus of the dCas9 protein to generate dCas9.TAD. The chimeric dCas9.TAD (Figure 7) fusion was generated by PCR amplifying the activation domain of the dHax3 (amino acid 683-960) and using MluI and XhoI restriction enzymes. For transcriptional repression, the SRDX motif was shown to be a potent and dominant repressor that can be fused to a variety of transcription factor for gene silencing purposes [94]. Therefore, we built a chimeric dCas9 repressor by fusing the

SRDX EAR motif (LDLDLELRLGFA) to the dCas9 C-terminus using MluI and
EcoRI restriction enzymes, to generate dCas9.SRDX (Figure 7).

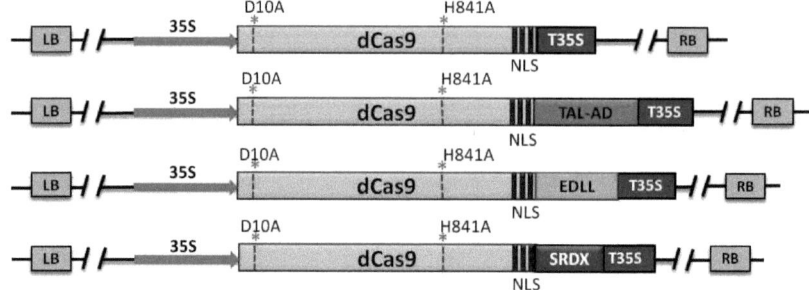

Figure 7: Schematic representation of dCas9 chimeric effector proteins. The catalytically inactive Cas9 harboring the
D10A and H841A nuclease domains mutations was cloned in pK2GW7 binary vector where the expression is driven
by 35S promoter and terminated by 35 terminator sequence. Only T-DNA region is depicted. The dCas9 backbone
was fused to dHax3 TAL activation domain, shown in red, to generate the dCas9:TAD synthetic transcriptional
activator. Similarly the EDLL activation domain, shown in green, was fused to the C-terminus of dCas9 to generate
dCas9.EDLL activator. For transcriptional repression, the SRDX domain, shown in purple, was fused to the C-
terminus of dCas9 to generate the dCas9.SRDX repressor.

To test the activity of the transcriptional activators, we constructed an episomal target
composed of Bs3 minimal promoter driving the uidA gene in pKGWFS7 vector
(Figure 8A). Since the targeting of the dCas9 protein requires the guide RNA
molecules, we constructed U6::gRNAs, which were subcloned in pYL156 vector (SI)
[96]. The gRNAs were designed to bind to several places in the Bs3 promoter on the
sense and antisense strands. To test the activity of our transcriptional regulators on an
endogenous target, we selected the *Nicotiana benthamiana* phytoene desaturase gene
(PDS). Thus, we generated few gRNAs capable of directing the dCas9 to several
positions in the promoter or first exon as shown in Figure 8B. The NGG PAM
sequence requirement, for the streptococcus pyogenes Cas9, was taken into
consideration in the designs of all targets. To test the functionality of our
transcriptional effectors we transformed all binary constructs of effectors, target and
gRNAs into Agrobacterium tumefaciens and co-delivered them, in different
combinations, into plant cells via agroinfiltration for transient assays (Figure 6).

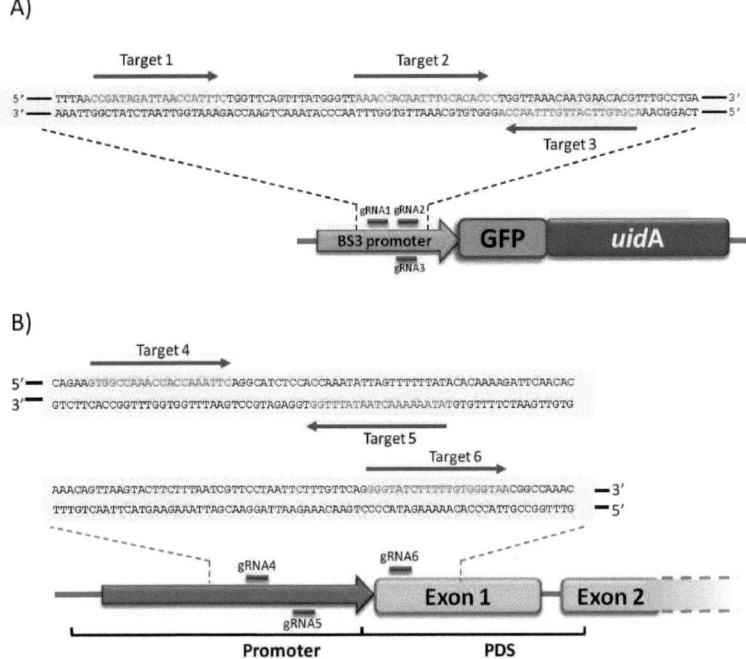

Figure 4: : Schematic representation of episomal and endogenous gRNA target sequences used in this study. A) Bs3 basal promoter driving the expression of the uidA in pKGWFS7 binary vector is used as an episomal target to test the activity of the synthetic transcriptional regulators on sense and antisense strands. gRNA target sequences are shown, selected on both sense and antisense strands preceding the PAM NGG sequences. B) PDS gene in *Nicotiana benthamiana* is used as an endogenous genomic target with different gRNA targets in the promoter region and the first exon.

3.2 The dCas9.TAs mediate strong transcriptional activation of a target gene.

We asked whether the chimeric dCas9.EDLL and dCas9.TAD transcriptional activators were guided by the gRNAs to the complementary sequence elements of the Bs3 promoter and whether they successfully activated the expression of the episomal *uidA* target gene (Figure 9A). To answer this question, we co-delivered the dCas9.TAs, Bs3::*uid*A episomal target and gRNAs in different combinations into three to four weeks old tobacco leaves *via Agrobacterium tumefaciens*. The gRNAs were targeted to different promoter elements on the sense and antisense strands of the

Bs3::uidA promoter. Infiltrated leaf disks were collected 36 to 48 hours post-infiltration and subjected to qualitative and quantitative GUS analysis. Our results indicate that the dCas9.EDLL synthetic transcriptional activator is capable of mediating site-specific transcriptional activation of the *uid*A reporter gene (Figure 9B). Several controls were used in this experiment including gRNA alone, *Bs3::uidA* target alone and gRNAs with non-complementary sequence to the Bs3 promoter. A minimal background activation was observed when Bs3::*uid*A was separately infiltrated into the tobacco leaves. Moreover, when dCas9.EDLL was co-infiltrated with gRNA6 which is non-complementary to the *Bs3* sequence, a pronounced level of basal background was noted. Although higher background was observed with the control gRNA, the target gRNAs mediated stronger and reproducible transcriptional activation in at least 5 independent experiments. To corroborate our transcriptional activation results with the dCas9.EDLL, we performed a similar set of experiment using the dCas9.TAD composed of the dCas9 and dhax3 TAL activation domain fusions. dCas9.TAD was guided to Bs3::*uid*A promoter by gRNA2 and was capable of mediating site-specific transcriptional activation, when compared to control combinations, indicating the versatility and reproducibility of the approach (Figure 9C). Next, we investigated whether dCas9 chimeric activators generated in this study are capable of enhancing expression levels of the endogenous *N. benthamiana* PDS gene. We used the chimeric dCas9.EDLL and dCas9.TAD proteins with several gRNA molecules designed to target the sense and antisense strands the PDS promoter as well as fist exon of PDS gene. RT-PCR gene expression analysis indicated an increase in the PDS expression level mediated by both dCas9.EDLL and dCas9.TAD transcriptional activators when compared to the controls. gRNAs targeting to the sense (gRNA4), antisense (gRNA5) and first exon (gRNA6) were capable of inducing comparable levels of activation of PDS gene. Moreover, delivering multiple gRNAs simultaneously did not trigger a synergistic effect on the activity level when compared to expression driven by single gRNA molecules.

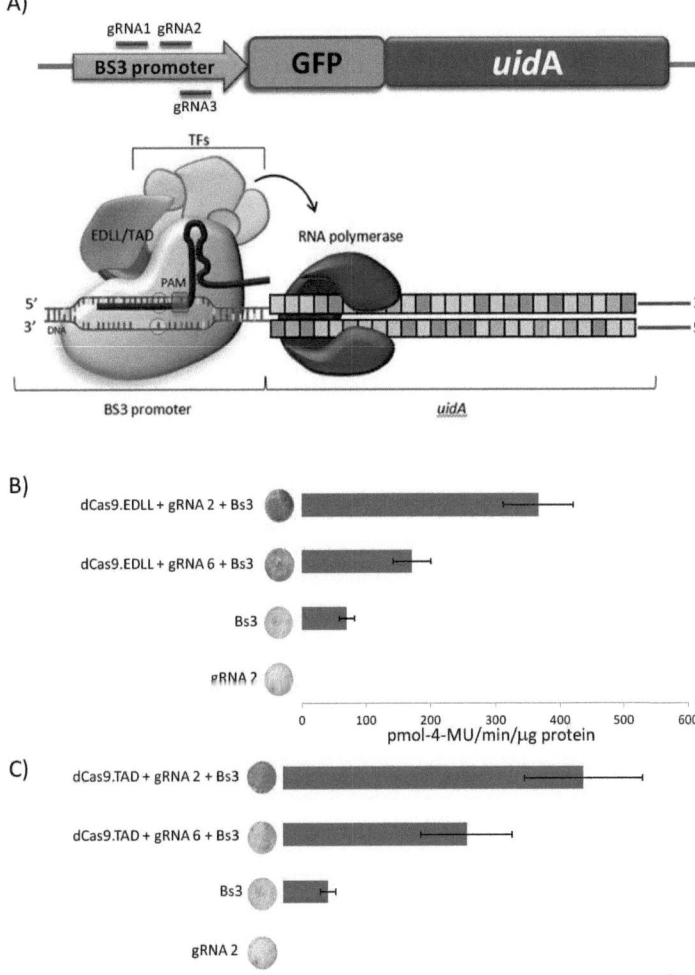

Figure 5: Transcriptional activation mediated by chimeric dCas9.TFs. A) Schematic representation of the episomal target used in this study and the assembly of the CRISPR/dCas9 complex on DNA. Whereas dCas9 mediates binding to the target DNA sequence, the activation domains recruit the transcription factors involved in the assembly and functioning of the transcriptional activation machinery. B) dCas9.EDLL activator strongly mediates transcriptional activation when guided to the Bs3 promoter compared to Bs3 control, and a sequence non-specific gRNA control C) Transcriptional activation using dCas9.TAD chimeric protein resulted in higher level of activation when compared to Bs3 control, and the sequence non-specific gRNA control. Error bars indicate standard error in the quantitative assay. Representative leaf disks of the qualitative assay are shown. These data represent 5 independent experiments, for each combination 3 to 4 plants were used.

3.3 Positional of gRNAs binding in the promoter influence transcriptional activation.

Since our previous experiment demonstrated that a successful transcriptional activation can be mediated by CRISPR/dCas9.TAs, we attempted to determine whether the transcriptional activation is influenced by the positions of the gRNA promoter binding sites relative to the transcriptional start site (TSS). Few gRNA molecules, complementary to the promoter elements and with variable lengths from the TSS were designed as follows: gRNA1: -297, gRNA2: -259 and gRNA3:-239. These gRNAs co-delivered with each transcriptional activator (dCas9.EDLL and dCas9.TAD), along with the Bs3::uidA in pKGWFS7 target vector. We determined the level of transcriptional activation mediated by dCas9.EDLL and dCas9.TAD by employing the Nicotiana benthamiana transient assays. Our results indicate that gRNA2 and gRNA1, targeting the sense strand, mediated higher transcriptional activation when compared with gRNA3, which targets the anti-sense strand, in the case of dCas9.EDLL (Figure 10A). In the dCas9.TAD experiments, activation of targets located further upstream of the transcription start site on the sense strand (gRNA1) are comparable with the targets located at the antisense strand (gRNA3). Additionally, gRNA2 mediates stronger activity level in comparison with gRNA1, indicating the importance of the gRNA binding distance from the TSS and how it might influence the transcriptional activity (Figure 10B). Further characterization is required using different promoters and with gRNAs binding sites that are widely spaced.

A)

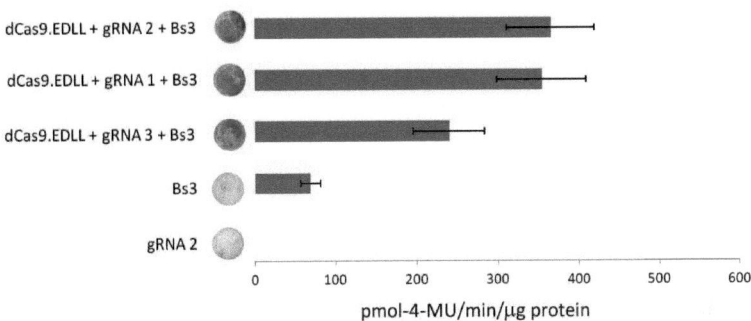

B)

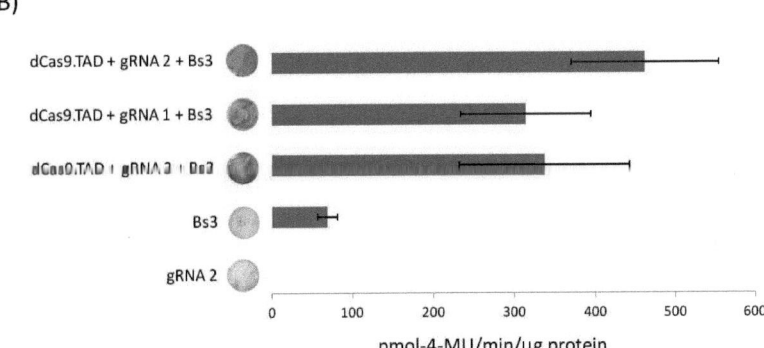

Figure 6: Positional effect of dCas9 promoter binding elements on transcriptional activation of *Bs3::uidA* episomal target gene. A) Positional effect of the dCas9.EDLL mediated transcriptional activation. gRNA1 and gRNA2 targeting sequences on the sense strand mediated stronger transcriptional activation than gRNA3 targeting the antisense strand when compared to the Bs3 control only. The delivery of the gRNA2 alone did not yield any positive GUS signal. B) Positional effect of the dCas9.TAD mediated transcriptional activation. gRNA2 targeting the sense strand closer to the TSS mediated the highest level of transcriptional activation. Notably, all gRNA targets mediated significantly higher level of activation in comparison with the Bs3 control. Error bars indicate standard error in the quantitative assay. Representative leaf disk of the qualitative assay is shown.

3.4 Transcriptional activation is enhanced by multiple gRNAs.

To test whether multiple gRNAs could result in synergistic effect on the transcriptional activation, we co-delivered dCas9.EDLL, Bs3::uidA episomal target and multiple gRNAs in different combinations into tobacco leaves. To measure the synergistic effect we used a combination of the gRNAs that bind to the sense and antisense strands on the episomal target with dCas9.EDLL and dCas9.TAD transcriptional activators. Our data show that using multiple gRNAs that bind to the sense strand increased the overall transcriptional activation when compared to the levels shown in single gRNA experiments. Using multiple gRNAs that bind to sense and antisense strand resulted in a lower level of transcriptional activation when compared with the levels observed using gRNAs targeting the sense strand.

In the dCas9.EDLL experiments the use of two gRNAs binding to the sense strand (gRNA2) and antisense strand (gRNA3) resulted in higher level of activation in comparison with the single gRNA experiment (Figure 11A). Moreover, we observed different transcriptional levels using different combination of gRNAs binding to the sense and antisense strand. For example, using gRNA2 and gRNA3 showed higher transcriptional activation level then using gRNA1 and gRNA3 (Figure 11A).

The same set of experiments was performed using dCas9.TAD. Combinations of gRNAs that bind to the sense and antisense strands have led to increased transcriptional activation levels when compared with gRNAs that bind to sense or antisense strand separately. Consistent with the data that were shown with dCas9.EDLL combination of the gRNAs targeting sense and antisense strand on locations closer to the transcription start site (gRNA2 and gRNA3) showed higher level of activation in comparison to combinations where the gRNAs targeting the sense strand were located further upstream (gRNA1 and gRNA3) (Figure 11B).

Our data indicate that the use of gRNAs binding to the sense strand increase the transcriptional activation levels of the uidA gene when compared to the other combinations. Further, the gRNA binding site and distance relative to the TSS in the promoter might influence the levels of transcriptional activation and thus further characterization and studies need to be performed for each promoter.

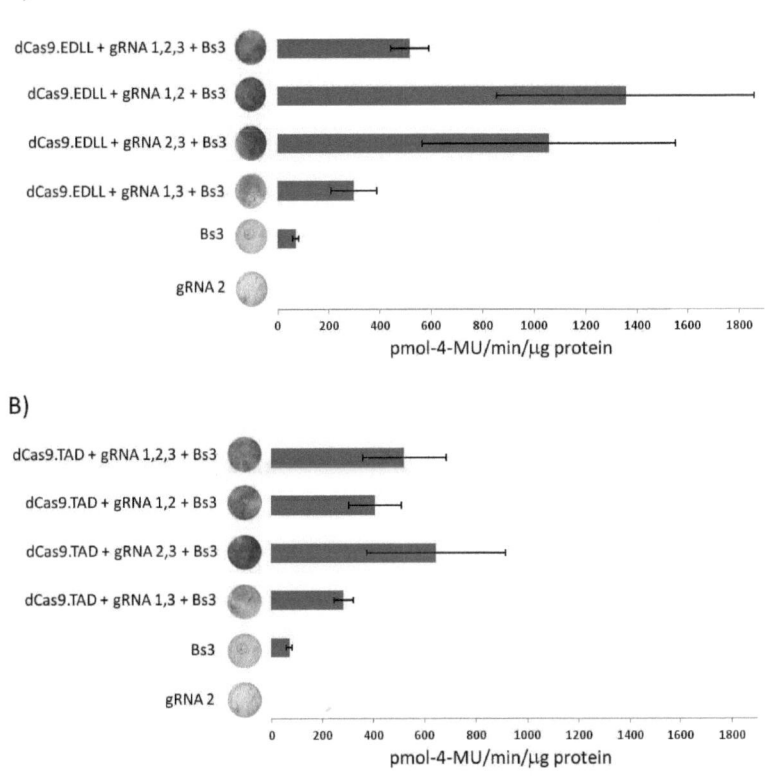

Figure 7: Synergistic effects of dCas9.TFs complexes on the overall transcriptional regulation of the Bs3 promoter. A) The highest activation level was observed using gRNA1 and gRNA2 targeting the sense strand in the Bs3 promoter. When combining targets on the sense and antisense strands, targets in proximal positions from each other (gRNA 2 and gRNA3) showed a higher level of activation meanwhile targets located further away from each other (gRNA1 and gRNA3) showed the lowest activation level. Delivery of all three gRNAs does not necessarily achieve the best level of activity. B) The highest activation level was achieved using gRNA2 and gRNA3 targeting sense and antisense strands respectively at proximal positions. In contrast gRNAs targeting the sense (gRNA1) and antisense (gRNA3) strands located further away from each other showed lowest activity level. The delivery of all three gRNAs together activated the GUS gene at a similar levels observed in the same combination used in the dCas9.EDLL experiment. Error bars indicate standard error in the quantitative assay. Representative leaf disk of the qualitative assay is shown.

3.5 The dCas9-SRDX mediates transcriptional repression of a PDS target gene.

CRISPR/dCas9 system has been shown to repress different target genes in bacterial and mammalian cells, respectively [75]. Gene repression effects were robust and reversible. In this study we attempted to test whether dCas9 fused to a repressor domain, such as SRDX (dCas9.SRDX), would enable site specific repression of plant genes (Figure 12A). Therefore, we constructed a chimeric dCas9.SRDX protein and several gRNA molecules designed to target and bind to the promoter region of phytoene desaturase gene (PDS) in N. benthamiana genome. Subsequently, we co-delivered via Agrobacterium tumefaciens the dCas9.SRDX effector proteins with the guide RNA molecules into three to four weeks old tobacco leaves. Infiltrated leaf samples were collected 36- to 48 hours post-infiltration (hpi) and the transcription regulation was assayed by semi-quantitative RT-PCR analysis. Since previous reports have shown that dCas9 alone is capable of repressing target genes, possibly by interfering with the transcriptional machinery or physical hindrance of RNA pol II (Figure 12A) [75,88,97], we attempted to test out whether dCas9 alone is capable of mediating such repression effects in planta. Interestingly, our data show that dCas9 guided to the first exon of the PDS gene (gRNA6) was capable of mildly repressing the PDS transcription. Moreover, dCas9 guided to the sense strand of the promoter region (gRNA4) did not change noticeably the PDS expression levels while dCas9 guided to the antisense strand resulted in elevated expression levels (Figure 12B). When dCas9.SRDX chimeric protein was used, a marked repression effects were demonstrated with all gRNAs targeting the exonic, sense and antisense strands of the promoter region (Figure 12B). Therefore, our data demonstrate that dCas9.SRDX transcriptional repressor is capable of mediating transcriptional repression of the PDS gene in a site-specific manner.

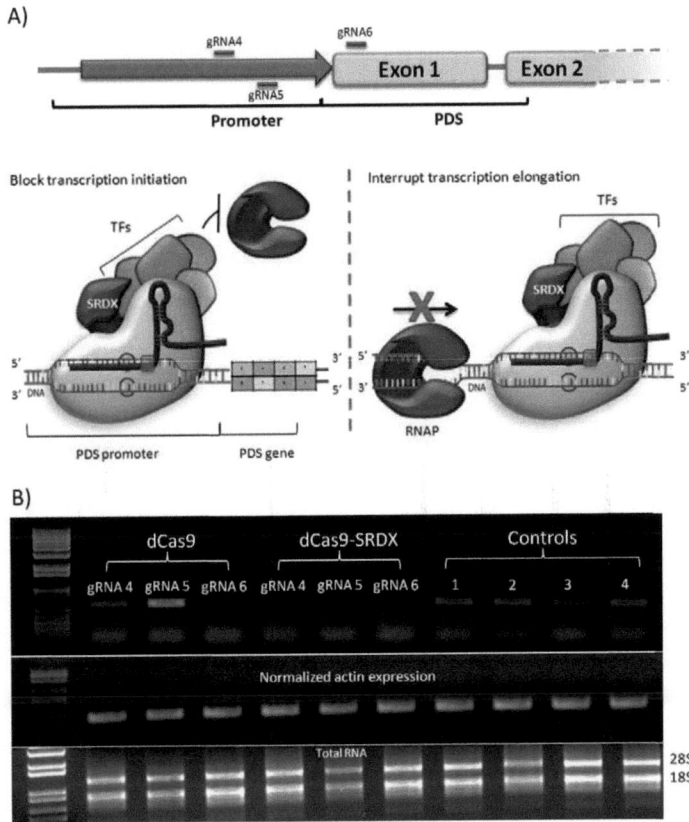

Figure 8: dCas9.SRDX mediated repression of PDS endogenous target in Nicotiana benthamiana transient leaf assays. A) gRNAs guide the dCas9 transcriptional repressor to the target site. Whereas dCas9 mediates binding to the target DNA sequence, SRDX repressor domain recruits the transcription factors responsible for blocking the expression of the gene. Repression can be achieved either by blocking the transcriptional initiation when dCas9.SRDX binds to the promoter region of the gene or by interrupting the transcriptional elongation when the target binding takes place downstream of the TSS. B) RT-PCR analysis of PDS gene expression level. (Lower panel) 18S and 28S RNA loading controls. (Middle panel) Normalized housekeeping actin1 gene expression. (Upper panel) Variable levels of transcriptional regulation of PDS gene. The expression pattern of PDS observed using dCas9 alone shows uneven repression levels. It seems that gRNA6 guiding the dCas9 to the first exon of the gene yields the strongest repression. The expression of the PDS gene is nearly completely down-regulated when targeting dCas9.SRDX to the sense (gRNA4) and antisense (gRNA5) strand in the PDS promoter, as well as dCas9.SRDX targeting the first exon (gRNA6) when compared with the positive controls. Control 1: dCas9 alone; control 2: dCas9.SRDX alone; control 3: dCas9+U6::gRNA (SI); control 4:dCas9.SRDX+U6::gRNA (SI).

3.6 The dCas9-EDLL and dCas9-TAD mediate transcriptional activation of a PDS target gene.

Next, we investigated whether the dCas9 chimeric activators generated in this study are capable of enhancing expression levels of the endogenous N. Benthamiana PDS gene. We used the chimeric dCas9.EDLL and dCas9.TAD proteins with several gRNA molecules designed to target the sense and antisense strands of the PDS promoter as well as the first exon of the PDS gene. RT-PCR gene expression analysis indicated an increase in the PDS expression level mediated by both dCas9.EDLL and dCas9.TAD transcriptional activators when compared to controls (Figure 13A). gRNAs targeting sense (gRNA4), antisense (gRNA5) and first exon (gRNA6) were capable of inducing comparable levels of activation of PDS gene (Figure 13B). Moreover, delivering multiple gRNAs simultaneously did not trigger a synergistic effect on the activity level when compared to the expression driven by single gRNA molecules (Figure13).

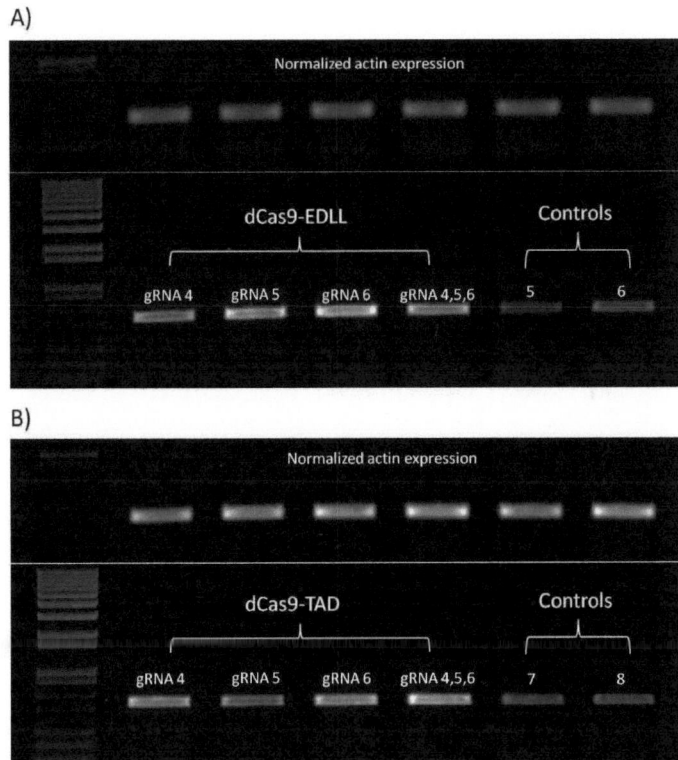

Figure 9: Transcriptional activation of endogenous PDS target using dCas9.TAs. A) RT-PCR analysis of PDS gene expression level mediated by dCas9.EDLL synthetic activator. (Upper panel) Normalized housekeeping actinI gene expression. (Lower panel) Similar activation levels were observed when using single gRNAs targeting the sense (gRNA4) and antisense (gRNA5) strands in the promoter region as well as gRNA6 targeting the sense strand of the first exon. Co-delivery of all three gRNAs does not seem to induce synergy. However all gRNA combinations show higher level of expression in comparison with the controls. Control 5: dCas9.EDLL; Control 6: dCas9.EDLL+U6::gRNA. B) RT-PCR analysis of PDS gene expression level mediated by dCas9.TAD synthetic activator. (upper panel) Normalized housekeeping actinI gene expression. (Lower panel) Using single and multiple combinations of gRNAs targeting PDS gene shows comparable levels of expression when compared to one another which exceeds the expression level of the controls. Control 7: dCas9.TAD; Control 8: dCas9.TAD+U6::gRNA.

3.7 Synthetic transcriptional repressor interferes and competes with transcriptional activation machinery.

It has been reported that SRDX domain is a dominant repressor that retains its repression function, even when fused to transcriptional factors containing activation domains [94,98-100]. We attempted to study the effect of dCas9.SRDX on transcriptional activation mediated by either dCas9.EDLL or dCas9.TAD chimeric transcriptional activators. Co-delivering repressor and activator constructs simultaneously, with single or multiple gRNAs targeting the Bs3 promoter of our episomal target, would enable studying the effects of the dCas9.SRDX repressor on the activity of the transcriptional activators. Since a single gRNA could guide either dCas9.SRDX or a dCas9.TA (dCas9.EDLL or dCas9.TAD) to the episomal target, this system would test the transcriptional interference of dCas9.SRDX with the activity of dCas9.TAs. For example, when dCas9.SRDX was co-delivered with dCas9.EDLL and gRNA2 combination, we observed significant reduction in the transcriptional activation levels normally mediated by dCas9.EDLL (Figure 14A). Similarly, dCas9.SRDX was capable of reducing the transcriptional activity mediated by dCas9.TAD (Figure 14B). Thus, dCas9.SRDX is capable of interfering with transcriptional activation mediated by either transcriptional activator.

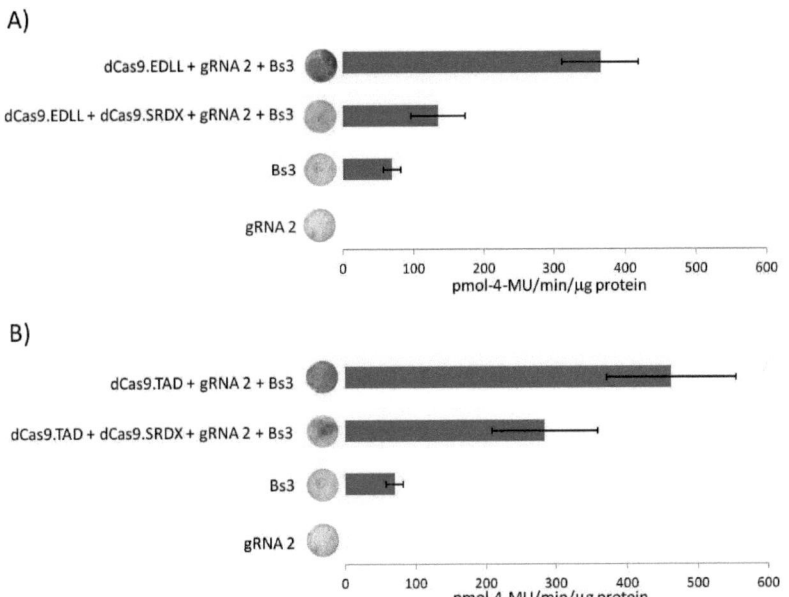

Figure 10: Interference analysis of the dCas9:SRDX repressor with dCas9:EDLL and dCas9:TAD transcriptional activators A). dCas9.SRDX is capable of interfering with the transcriptional activation mediated by dCas9:EDLL . High expression level observed using dCas9.EDLL decreased dramatically when dCas9.SRDX repressor co-delivered with dCas9.EDLL.. B) dCas9.SRDX is capable of mildly interfering with the transcriptional activation mediated by dCas9:TAD.

4 Discussion:

In the present work, we show that the CRISPR/dCas9 system can be used as a platform for targeted genome regulation on episomal and genomic targets (Figure 7). We employed the catalytically inactive Cas9 protein combined with gRNA molecules as a DNA targeting module. We optimized the CRISPR/dCas9 system for in planta expression by subcloning the dCas9 gene into a binary vector under 35S constitutive promoter (pK2GW7). The gRNA molecules were subcloned downstream of Pol III U6 promoter into the pYL156 vector. The two binary vectors carrying the dCas9 protein and its gRNA were co-delivered into plant cells via Agrobacterium tumefaciens to test their activities on either episomal or genomic targets (Figure 6). It has been shown that the CRISPR/dCas9 mediates DNA targeting in bacterial and human cells [75,101]. Moreover, chimeric proteins generated by domain fusions to the dCas9 protein were capable of mediating targeted transcriptional regulation [102]. Since dCas9 was shown to function as a transcriptional repressor by interfering with the formation of the transcriptional machinery complex or hindering the RNA Pol II activity [75], we attempted to test whether this CRSIPR/dCas9 ribonucleocomplex could mediate DNA targeting and transcriptional activation or repression in planta. To do this, we generated two dCas9 chimeric transcriptional activators by fusing different transcriptional activation domains to the C-terminus of dCas9 (Figure 7). The EDLL domain of ERF/EREBP family of plant transcriptional factors was used because it has been established to be a strong transcriptional activator, easily transferable to other proteins and can bind to either proximal or distal positions from the TATA box [93]. The second dCas9 transcriptional activator was generated by fusing the activation domain of the dHax3 TAL effector protein of phytopathogenic, Xanthomonas sp, bacteria, due to its strong and well defined nature of activation of the plant host target genes. We used Bs3 promoter of the resistance (R) gene from the dicot pepper plants [103], fused with uidA gene as an episomal target. We tested the transcriptional activation and estimated the activation levels qualitatively and quantitatively, through GUS expression assays.

Several gRNAs complementary to either the sense or antisense strands in the Bs3 promoter region were generated in pYL156 binary vector under the control of U6 promoter. Effector and guide RNA fused to target constructs were co-delivered simultaneously into Nicotiana benthamiana leaves for transient assay.

Qualitative and quantitative analysis of the uidA gene indicate that the dCas9.EDLL fusion protein is capable of activating transcription by 5 folds when compared to the Bs3 control alone (Figure 9B). Similarly the dCas9.TAD was capable by mediating the transcriptional activation by 6 folds (Figure 9C). When using a control that targeted a sequence outside the Bs3 promoter we observed a level of activation of 2-folds and 4-folds in dCas9.EDLL and dCas9.TAD respectively (Figure 9B,C). This is not surprising and can be explained by the fact that a high copy number of the episomal molecules will be screened by the CRISPR/dCas9.TFs machinery to find gRNA complementary sequences [104-106]. Thus, the process of binding and screening for the complementary target could mildly activate the expression. Our interpretation of these data requires further characterization. Overall, our data demonstrate the successful targeting and transcriptional regulation as evident by the activation of uidA. Next, we attempted to study whether gRNA molecules targeted to the antisense strand could bring about the same activation effects as those targeting the sense strand. We found that transcriptional activation from gRNAs targeting the sense strands is 1.5 folds higher than those targeting the antisense strand possibly due to a structural flexibility in bringing in the transcriptional machinery and RNA polymerase activity (Figure 10). Moreover, we tested whether the distance from the transcriptional start site (TSS) could play a role in the transcriptional activation. We used several gRNAs targeted to -297 (gRNA1), -259 (gRNA2) and -239 (gRNA3) from the TSS and compared the level of transcriptional activity induced by each of them (Figure 10). With both transcriptional activators, grNA2, binding closer to the TSS shows higher activity than grNA1. This difference is significantly observed with dCas9.TAD activation domain in contrast to the slight difference observed using dCas9.EDLL. These data indicate that variable dCas9.Tas would require different distances from the TSS for optimum and higher transcriptional levels due primarily to structural requirements for the chimeric dCas9 fusions.

Furthermore, we wanted to determine whether there is an additive or synergistic effect by using multiple grNAs simultaneously targeted to sense and antisense strands (Figure 11).

We observed strong activation levels when we co-delivered two grNAs targeting the sense strand (grNA1 and grNA2) when compared to any other combination using the dCas9.EDLL (Figure 11A). These data indicate that such combination can be used to produce higher transcriptional levels of user-selected genomic targets. Interestingly, it is not the case when using dCas9.TAD (Figure 11B). The strongest activation was observed when mixing the grNA2 targeting the sense strand closer to the TSS and grNA3 targeting the antisense strand. This could indicate that the dCas9.TAD complexes binding could be feasible on sense and antisense strands simultaneously. Interestingly, the weakest additive transcriptional activation effect is achieved when co-delivering grNAs targeting the sense strand farther to the TTS, and antisense strand. These data reveal that using grNAs targeted to the antisense strand of the promoter might compromise the additive effects possibly due to the structural hindrance to the RNA pol II binding and transcriptional initiation machinery. Our data indicate that using multiple grNAs and assembling transcriptional complexes on different promoter regions lead to an increase in the overall transcriptional activation. Furthermore, our data indicate that it may not be necessary to use several grNA molecules simultaneously to obtain the most robust transcriptional effect and that the number of grNA molecules and dCas9 complexes may reach a saturation point where using more gRNAs might structurally hinder the assembly and activity of neighboring complexes and decrease the overall intended transcriptional activation effect.

Since we demonstrated the transcriptional activation on episomal targets, we argued that it might be possible to generate transcriptional repressors capable of down regulating genomic targets in a site-specific manner. Therefore, we fused a dominant repression domain, SRDX, from the ERF transcription factors family [107,108] to the C-terminus of the dCas9 protein (Figure 7). This domain has been used before to down regulate target genes when fused to transcription factors or TAL effectors as DNA binding modules [100]. We delivered the dCas9.SRDX synthetic repressor with gRNAs targeting different parts of the promoter of the PDS gene of *Nicotiana benthamiana* (Figure 8B). Our data show that the dCas9.SRDX synthetic repressor is capable of mediating transcription repression of the target genes as shown in Figure 12B. In conclusion, our present study provides the first evidence that CRSIPSR/dCas9 can be used for targeted plant genome regulation. Moreover, we built several synthetic transcriptional factors using activation and repression domains and demonstrated the feasibility of targeted transcriptional regulation on episomal as well as genomic targets in planta. Several areas of improvements still remain including using shorter versions of dCas9 capable of mediating DNA targeting and optimizing strong repression and activation domains to the dCas9 improved architecture. Generation of dCas9.TF over expression lines, for example in Arabidopsis and tobacco, could facilitate the functional studies of single and multiple genes under different nutritional, hormonal or environmental treatments respectively. Because gRNA molecules can be delivered using weak symptomless virus capable of systemic infection of the plant, using such approach it might be possible to build a functional genomics platform in multiple plants species amenable to symptomless viral infection as gRNA carriers. We can envision that a trait discovery platform can be built by regulating the delivery of multiple gRNAs targeting single, multiple or all members of a gene family. Although the current study awaits further characterization and application on stable transgenic lines over-expressing dCas9, the present study opens enormous possibilities to address important basic plant biology questions and enhance agricultural biotechnology applications.

REFERENCES

1. Kakimoto T: CKI1, a histidine kinase homolog implicated in cytokinin signal transduction. *Science* 1996, 274:982-985.

2. Weigel D, Ahn JH, Blazquez MA, Borevitz JO, Christensen SK, Fankhauser C, Ferrandiz C, Kardailsky I, Malancharuvil EJ, Neff MM, et al.: Activation tagging in Arabidopsis. *Plant Physiol* 2000, 122:1003-1013.

3. Ichikawa T, Nakazawa M, Kawashima M, Iizumi H, Kuroda H, Kondou Y, Tsuhara Y, Suzuki K, Ishikawa A, Seki M, et al.: The FOX hunting system: an alternative gain-of-function gene hunting technique. *Plant J* 2006, 48:974-985.

4. Napoli C, Lemieux C, Jorgensen R: Introduction of a Chimeric Chalcone Synthase Gene into Petunia Results in Reversible Co-Suppression of Homologous Genes in Trans. *Plant Cell* 1990, 2:279-289.

5. Pal-Bhadra M, Bhadra U, Birchler JA: Cosuppression in Drosophila: gene silencing of Alcohol dehydrogenase by white-Adh transgenes is Polycomb dependent. *Cell* 1997, 90:479-490.

6. Vaucheret H, Beclin C, Fagard M: Post-transcriptional gene silencing in plants. *J Cell Sci* 2001, 114:3083-3091.

7. Cameron FH, Jennings PA: Inhibition of gene expression by a short sense fragment. *Nucleic Acids Res* 1991, 19:469-475.

8. de Carvalho Niebel F, Frendo P, Van Montagu M, Cornelissen M: Post-transcriptional cosuppression of beta-1,3-glucanase genes does not affect accumulation of transgene nuclear mRNA. *Plant Cell* 1995, 7:347-358.

9. Dorlhac de Borne F, Vincentz M, Chupeau Y, Vaucheret H: Co-suppression of nitrate reductase host genes and transgenes in transgenic tobacco plants. *Mol Gen Genet* 1994, 243:613-621.

10. Vaucheret H, Nussaume L, Palauqui JC, Quillere I, Elmayan T: A Transcriptionally Active State Is Required for Post-Transcriptional Silencing

(Cosuppression) of Nitrate Reductase Host Genes and Transgenes. *Plant Cell* 1997, 9:1495-1504.

11. Hamilton AJ, Baulcombe DC: A species of small antisense RNA in posttranscriptional gene silencing in plants. *Science* 1999, 286:950-952.

12. Boyd AL, Samid D: Review: molecular biology of transgenic animals. *J Anim Sci* 1993, 71 Suppl 3:1-9.

13. Guo S, Kemphues KJ: par-1, a gene required for establishing polarity in C. elegans embryos, encodes a putative Ser/Thr kinase that is asymmetrically distributed. *Cell* 1995, 81:611-620.

14. Martinez J, Patkaniowska A, Urlaub H, Luhrmann R, Tuschl T: Single-stranded antisense siRNAs guide target RNA cleavage in RNAi. *Cell* 2002, 110:563-574.

15. Kennedy S, Wang D, Ruvkun G: A conserved siRNA-degrading RNase negatively regulates RNA interference in C. elegans. *Nature* 2004, 427:645-649.

16. Sharp PA: RNA interference--2001. *Genes Dev* 2001, 15:485-490.

17. Lipardi C, Wei Q, Paterson BM: RNAi as random degradative PCR: siRNA primers convert mRNA into dsRNAs that are degraded to generate new siRNAs. *Cell* 2001, 107:297-307.

18. Bernstein E, Caudy AA, Hammond SM, Hannon GJ: Role for a bidentate ribonuclease in the initiation step of RNA interference. *Nature* 2001, 409:363-366.

19. Hammond SM, Bernstein E, Beach D, Hannon GJ: An RNA-directed nuclease mediates post-transcriptional gene silencing in Drosophila cells. *Nature* 2000, 404:293-296.

20. Porteus MH, Carroll D: Gene targeting using zinc finger nucleases. *Nat Biotech* 2005, 23:967-973.

21. Bibikova M, Beumer K, Trautman JK, Carroll D: Enhancing Gene Targeting with Designed Zinc Finger Nucleases. *Science* 2003, 300:764.

22. Porteus MH, Baltimore D: Chimeric nucleases stimulate gene targeting in human cells. *Science* 2003, 300:763.

23. Urnov FD, Rebar EJ, Holmes MC, Zhang HS, Gregory PD: Genome editing with engineered zinc finger nucleases. *Nat Rev Genet* 2010, 11:636-646.

24. Miller JC, Holmes MC, Wang J, Guschin DY, Lee YL, Rupniewski I, Beausejour CM, Waite AJ, Wang NS, Kim KA, et al.: An improved zinc-finger nuclease architecture for highly specific genome editing. *Nat Biotechnol* 2007, 25:778-785.

25. Liu Q, Segal DJ, Ghiara JB, Barbas CF, 3rd: Design of polydactyl zinc-finger proteins for unique addressing within complex genomes. *Proc Natl Acad Sci U S A* 1997, 94:5525-5530.

26. Beerli RR, Dreier B, Barbas CF, 3rd: Positive and negative regulation of endogenous genes by designed transcription factors. *Proc Natl Acad Sci U S A* 2000, 97:1495-1500.

27. Beerli RR, Segal DJ, Dreier B, Barbas CF, 3rd: Toward controlling gene expression at will: specific regulation of the erbB-2/HER-2 promoter by using polydactyl zinc finger proteins constructed from modular building blocks. *Proc Natl Acad Sci U S A* 1998, 95:14628-14633.

28. Elrod-Erickson M, Rould MA, Nekludova L, Pabo CO: Zif268 protein-DNA complex refined at 1.6 A: a model system for understanding zinc finger-DNA interactions. *Structure* 1996, 4:1171-1180.

29. Ptashne M: 1997 Albert Lasker Award for Basic Medical Research. Control of gene transcription: an outline. *Nat Med* 1997, 3:1069-1072.

30. Estruch JJ, Crossland L, Goff SA: Plant activating sequences: positively charged peptides are functional as transcriptional activation domains. *Nucleic Acids Res* 1994, 22:3983-3989.

31. Hall DB, Struhl K: The VP16 activation domain interacts with multiple transcriptional components as determined by protein-protein cross-linking in vivo. *J Biol Chem* 2002, 277:46043-46050.

32. Stege JT, Guan X, Ho T, Beachy RN, Barbas CF, 3rd: Controlling gene expression in plants using synthetic zinc finger transcription factors. *Plant J* 2002, 32:1077-1086.

33. Beerli RR, Barbas CF, 3rd: Engineering polydactyl zinc-finger transcription factors. *Nat Biotechnol* 2002, 20:135-141.

34. Margolin JF, Friedman JR, Meyer WK, Vissing H, Thiesen HJ, Rauscher FJ, 3rd: Kruppel-associated boxes are potent transcriptional repression domains. *Proc Natl Acad Sci U S A* 1994, 91:4509-4513.

35. Heinzel T, Lavinsky RM, Mullen TM, Soderstrom M, Laherty CD, Torchia J, Yang WM, Brard G, Ngo SD, Davie JR, et al.: A complex containing N-CoR, mSin3 and histone deacetylase mediates transcriptional repression. *Nature* 1997, 387:43-48.

36. Gersbach CA, Gaj T, Gordley RM, Mercer AC, Barbas CF, 3rd: Targeted plasmid integration into the human genome by an engineered zinc-finger recombinase. *Nucleic Acids Res* 2011, 39:7868-7878.

37. Gaj T, Mercer AC, Gersbach CA, Gordley RM, Barbas CF, 3rd: Structure-guided reprogramming of serine recombinase DNA sequence specificity. *Proc Natl Acad Sci U S A* 2011, 108:498-503.

38. Xu GL, Bestor TH: Cytosine methylation targetted to pre-determined sequences. *Nat Genet* 1997, 17:376-378.

39. McNamara AR, Hurd PJ, Smith AE, Ford KG: Characterisation of site-biased DNA methyltransferases: specificity, affinity and subsite relationships. *Nucleic Acids Res* 2002, 30:3818-3830.

40. Yant SR, Huang Y, Akache B, Kay MA: Site-directed transposon integration in human cells. *Nucleic Acids Res* 2007, 35:e50.

41. Smith MC, Thorpe HM: Diversity in the serine recombinases. *Mol Microbiol* 2002, 44:299-307.

42. Grindley ND, Whiteson KL, Rice PA: Mechanisms of site-specific recombination. *Annu Rev Biochem* 2006, 75:567-605.

43. Razin A: CpG methylation, chromatin structure and gene silencing-a three-way connection. *EMBO J* 1998, 17:4905-4908.

44. Ng HH, Bird A: DNA methylation and chromatin modification. *Curr Opin Genet Dev* 1999, 9:158-163.

45. Wyszynski MW, Gabbara S, Bhagwat AS: Substitutions of a cysteine conserved among DNA cytosine methylases result in a variety of phenotypes. *Nucleic Acids Res* 1992, 20:319-326.

46. Mi S, Roberts RJ: The DNA binding affinity of HhaI methylase is increased by a single amino acid substitution in the catalytic center. *Nucleic Acids Res* 1993, 21:2459-2464.

47. Durai S, Mani M, Kandavelou K, Wu J, Porteus MH, Chandrasegaran S: Zinc finger nucleases: custom-designed molecular scissors for genome engineering of plant and mammalian cells. *Nucleic Acids Research* 2005, 33:5978-5990.

48. Wolfe SA, Nekludova L, Pabo CO: DNA recognition by Cys2His2 zinc finger proteins. *Annu Rev Biophys Biomol Struct* 2000, 29:183-212.

49. Pattanayak V, Ramirez CL, Joung JK, Liu DR: Revealing off-target cleavage specificities of zinc-finger nucleases by in vitro selection. *Nat Meth* 2011, 8:765-770.

50. Cathomen T, Keith Joung J: Zinc-finger Nucleases: The Next Generation Emerges. *Mol Ther* 2008, 16:1200-1207.

51. Boch J, Scholze H, Schornack S, Landgraf A, Hahn S, Kay S, Lahaye T, Nickstadt A, Bonas U: Breaking the code of DNA binding specificity of TAL-type III effectors. *Science* 2009, 326:1509-1512.

52. Boch J, Bonas U: Xanthomonas AvrBs3 family-type III effectors: discovery and function. *Annu Rev Phytopathol* 2010, 48:419-436.

53. Romer P, Hahn S, Jordan T, Strauss T, Bonas U, Lahaye T: Plant pathogen recognition mediated by promoter activation of the pepper Bs3 resistance gene. *Science* 2007, 318:645-648.

54. Zheng CK, Wang CL, Zhang XP, Wang FJ, Qin TF, Zhao KJ: The last half-repeat of transcription activator-like effector is dispensable and thereby TALE-based technology can be simplified. *Mol Plant Pathol* 2014.

55. Moscou MJ, Bogdanove AJ: A Simple Cipher Governs DNA Recognition by TAL Effectors. *Science* 2009, 326:1501-1501.

56. Streubel J, Blucher C, Landgraf A, Boch J: TAL effector RVD specificities and efficiencies. *Nat Biotechnol* 2012, 30:593-595.

57. Mak AN, Bradley P, Cernadas RA, Bogdanove AJ, Stoddard BL: The crystal structure of TAL effector PthXo1 bound to its DNA target. *Science* 2012, 335:716-719.

58. Deng D, Yan C, Pan X, Mahfouz M, Wang J, Zhu JK, Shi Y, Yan N: Structural basis for sequence-specific recognition of DNA by TAL effectors. *Science* 2012, 335:720-723.

59. Morbitzer R, Romer P, Boch J, Lahaye T: Regulation of selected genome loci using de novo-engineered transcription activator-like effector (TALE)-type transcription factors. *Proc Natl Acad Sci U S A* 2010, 107:21617-21622.

60. Mahfouz MM, Li L, Piatek M, Fang X, Mansour H, Bangarusamy DK, Zhu JK: Targeted transcriptional repression using a chimeric TALE-SRDX repressor protein. *Plant Mol Biol* 2012, 78:311-321.

61. Juillerat A, Dubois G, Valton J, Thomas S, Stella S, Marechal A, Langevin S, Benomari N, Bertonati C, Silva GH, et al.: Comprehensive analysis of the specificity of transcription activator-like effector nucleases. *Nucleic Acids Res* 2014.

62. Makarova KS, Haft DH, Barrangou R, Brouns SJ, Charpentier E, Horvath P, Moineau S, Mojica FJ, Wolf YI, Yakunin AF, et al.: Evolution and classification of the CRISPR-Cas systems. *Nat Rev Microbiol* 2011, 9:467-477.

63. Deltcheva E, Chylinski K, Sharma CM, Gonzales K, Chao Y, Pirzada ZA, Eckert MR, Vogel J, Charpentier E: CRISPR RNA maturation by trans-encoded small RNA and host factor RNase III. *Nature* 2011, 471:602-607.

64. Gesner EM, Schellenberg MJ, Garside EL, George MM, Macmillan AM: Recognition and maturation of effector RNAs in a CRISPR interference pathway. *Nat Struct Mol Biol* 2011, 18:688-692.

65. Wiedenheft B, Sternberg SH, Doudna JA: RNA-guided genetic silencing systems in bacteria and archaea. *Nature* 2012, 482:331-338.

66. Bhaya D, Davison M, Barrangou R: CRISPR-Cas systems in bacteria and archaea: versatile small RNAs for adaptive defense and regulation. *Annu Rev Genet* 2011, 45:273-297.

67. Terns MP, Terns RM: CRISPR-based adaptive immune systems. *Curr Opin Microbiol* 2011, 14:321-327.

68. Sun J, Jeon JII, Shin M, Shin HC, Oh BH, Kim IS: Crystal structure and CRISPR RNA-binding site of the Cmr1 subunit of the Cmr interference complex. *Acta Crystallogr D Biol Crystallogr* 2014, 70:535-543.

69. Cong L, Ran FA, Cox D, Lin S, Barretto R, Habib N, Hsu PD, Wu X, Jiang W, Marraffini LA, et al.: Multiplex genome engineering using CRISPR/Cas systems. *Science* 2013, 339:819-823.

70. Jinek M, Chylinski K, Fonfara I, Hauer M, Doudna JA, Charpentier E: A programmable dual-RNA-guided DNA endonuclease in adaptive bacterial immunity. *Science* 2012, 337:816-821.

71. Nishimasu H, Ran FA, Hsu PD, Konermann S, Shehata SI, Dohmae N, Ishitani R, Zhang F, Nureki O: Crystal Structure of Cas9 in Complex with Guide RNA and Target DNA. *Cell* 2014.

72. Sternberg SH, Redding S, Jinek M, Greene EC, Doudna JA: DNA interrogation by the CRISPR RNA-guided endonuclease Cas9. *Nature* 2014.

73. Ran FA, Hsu PD, Lin CY, Gootenberg JS, Konermann S, Trevino AE, Scott DA, Inoue A, Matoba S, Zhang Y, et al.: Double nicking by RNA-guided CRISPR Cas9 for enhanced genome editing specificity. *Cell* 2013, 154:1380-1389.

74. Puchta H: Gene replacement by homologous recombination in plants. *Plant Mol Biol* 2002, 48:173-182.

75. Qi LS, Larson MH, Gilbert LA, Doudna JA, Weissman JS, Arkin AP, Lim WA: Repurposing CRISPR as an RNA-guided platform for sequence-specific control of gene expression. *Cell* 2013, 152:1173-1183.

76. DiCarlo JE, Norville JE, Mali P, Rios X, Aach J, Church GM: Genome engineering in Saccharomyces cerevisiae using CRISPR-Cas systems. *Nucleic Acids Res* 2013, 41:4336-4343.

77. Li JF, Norville JE, Aach J, McCormack M, Zhang D, Bush J, Church GM, Sheen J: Multiplex and homologous recombination-mediated genome editing in Arabidopsis and Nicotiana benthamiana using guide RNA and Cas9. *Nat Biotechnol* 2013, 31:688-691.

78. Nekrasov V, Staskawicz B, Weigel D, Jones JD, Kamoun S: Targeted mutagenesis in the model plant Nicotiana benthamiana using Cas9 RNA-guided endonuclease. *Nat Biotechnol* 2013, 31:691-693.

79. Shan Q, Wang Y, Li J, Zhang Y, Chen K, Liang Z, Zhang K, Liu J, Xi JJ, Qiu JL, et al.: Targeted genome modification of crop plants using a CRISPR-Cas system. *Nat Biotechnol* 2013, 31:686-688.

80. Wang H, Yang H, Shivalila CS, Dawlaty MM, Cheng AW, Zhang F, Jaenisch R: One-step generation of mice carrying mutations in multiple genes by CRISPR/Cas-mediated genome engineering. *Cell* 2013, 153:910-918.

81. Shen B, Zhang J, Wu H, Wang J, Ma K, Li Z, Zhang X, Zhang P, Huang X: Generation of gene-modified mice via Cas9/RNA-mediated gene targeting. *Cell Res* 2013, 23:720-723.

82. Li D, Qiu Z, Shao Y, Chen Y, Guan Y, Liu M, Li Y, Gao N, Wang L, Lu X, et al.: Heritable gene targeting in the mouse and rat using a CRISPR-Cas system. *Nat Biotechnol* 2013, 31:681-683.

83. Yang D, Xu J, Zhu T, Fan J, Lai L, Zhang J, Chen YE: Effective gene targeting in rabbits using RNA-guided Cas9 nucleases. *J Mol Cell Biol* 2014, 6:97-99.

84. Nakayama T, Fish MB, Fisher M, Oomen-Hajagos J, Thomsen GH, Grainger RM: Simple and efficient CRISPR/Cas9-mediated targeted mutagenesis in Xenopus tropicalis. *Genesis* 2013, 51:835-843.

85. Yu Z, Ren M, Wang Z, Zhang B, Rong YS, Jiao R, Gao G: Highly efficient genome modifications mediated by CRISPR/Cas9 in Drosophila. *Genetics* 2013, 195:289-291.

86. Bassett AR, Tibbit C, Ponting CP, Liu JL: Highly efficient targeted mutagenesis of Drosophila with the CRISPR/Cas9 system. *Cell Rep* 2013, 4:220-228.

87. Whelan WL, Gocke E, Manney TR: The CAN1 locus of Saccharomyces cerevisiae: fine-structure analysis and forward mutation rates. *Genetics* 1979, 91:35-51.

88. Gilbert LA, Larson MH, Morsut L, Liu Z, Brar GA, Torres SE, Stern-Ginossar N, Brandman O, Whitehead EH, Doudna JA, et al.: CRISPR-mediated modular RNA-guided regulation of transcription in eukaryotes. *Cell* 2013, 154:442-451.

89. Cheng AW, Wang H, Yang H, Shi L, Katz Y, Theunissen TW, Rangarajan S, Shivalila CS, Dadon DB, Jaenisch R: Multiplexed activation of endogenous genes by CRISPR-on, an RNA-guided transcriptional activator system. *Cell Res* 2013, 23:1163-1171.

90. Fukao T, Bailey-Serres J: Submergence tolerance conferred by Sub1A is mediated by SLR1 and SLRL1 restriction of gibberellin responses in rice. *Proc Natl Acad Sci U S A* 2008, 105:16814-16819.

91. Ohme-Takagi M, Shinshi H: Ethylene-inducible DNA binding proteins that interact with an ethylene-responsive element. *Plant Cell* 1995, 7:173-182.

92. Sakuma Y, Liu Q, Dubouzet JG, Abe H, Shinozaki K, Yamaguchi-Shinozaki K: DNA-binding specificity of the ERF/AP2 domain of Arabidopsis DREBs, transcription factors involved in dehydration- and cold-inducible gene expression. *Biochem Biophys Res Commun* 2002, 290:998-1009.

93. Tiwari SB, Belachew A, Ma SF, Young M, Ade J, Shen Y, Marion CM, Holtan HE, Bailey A, Stone JK, et al.: The EDLL motif: a potent plant transcriptional activation domain from AP2/ERF transcription factors. *Plant J* 2012, 70:855-865.

94. Hiratsu K, Matsui K, Koyama T, Ohme-Takagi M: Dominant repression of target genes by chimeric repressors that include the EAR motif, a repression domain, in Arabidopsis. *Plant J* 2003, 34:733-739.

95. Hiratsu K, Ohta M, Matsui K, Ohme-Takagi M: The SUPERMAN protein is an active repressor whose carboxy-terminal repression domain is required for the development of normal flowers. *FEBS Lett* 2002, 514:351-354.

96. Burch-Smith TM, Schiff M, Liu Y, Dinesh-Kumar SP: Efficient virus-induced gene silencing in Arabidopsis. *Plant Physiol* 2006, 142:21-27.

97. Larson MH, Gilbert LA, Wang X, Lim WA, Weissman JS, Qi LS: CRISPR interference (CRISPRi) for sequence-specific control of gene expression. *Nat Protoc* 2013, 8:2180-2196.

98. Heyl A, Ramireddy E, Brenner WG, Riefler M, Allemeersch J, Schmulling T: The transcriptional repressor ARR1-SRDX suppresses pleiotropic cytokinin activities in Arabidopsis. *Plant Physiol* 2008, 147:1380-1395.

99. Takada S: Post-embryonic induction of ATML1-SRDX alters the morphology of seedlings. *PLoS One* 2013, 8:e79312.

100. Mahfouz MM, Li LX, Piatek M, Fang XY, Mansour H, Bangarusamy DK, Zhu JK: Targeted transcriptional repression using a chimeric TALE-SRDX repressor protein. *Plant Molecular Biology* 2012, 78:311-321.

101. Bikard D, Jiang W, Samai P, Hochschild A, Zhang F, Marraffini LA: Programmable repression and activation of bacterial gene expression using an engineered CRISPR-Cas system. *Nucleic Acids Res* 2013, 41:7429-7437.

102. Perez-Pinera P, Kocak DD, Vockley CM, Adler AF, Kabadi AM, Polstein LR, Thakore PI, Glass KA, Ousterout DG, Leong KW, et al.: RNA-guided gene activation by CRISPR-Cas9-based transcription factors. *Nat Methods* 2013, 10:973-976.

103. Romer P, Recht S, Lahaye T: A single plant resistance gene promoter engineered to recognize multiple TAL effectors from disparate pathogens. *Proc Natl Acad Sci U S A* 2009, 106:20526-20531.

104. Jinek M, Jiang F, Taylor DW, Sternberg SH, Kaya E, Ma E, Anders C, Hauer M, Zhou K, Lin S, et al.: Structures of Cas9 endonucleases reveal RNA-mediated conformational activation. *Science* 2014, 343:1247997.

105. Nishimasu H, Ran FA, Hsu PD, Konermann S, Shehata SI, Dohmae N, Ishitani R, Zhang F, Nureki O: Crystal structure of Cas9 in complex with guide RNA and target DNA. *Cell* 2014, 156:935-949.

106. Sternberg SH, Redding S, Jinek M, Greene EC, Doudna JA: DNA interrogation by the CRISPR RNA-guided endonuclease Cas9. *Nature* 2014, 507:62-67.

107. Kagale S, Rozwadowski K: EAR motif-mediated transcriptional repression in plants: an underlying mechanism for epigenetic regulation of gene expression. *Epigenetics* 2011, 6:141-146.

108. Ohta M, Matsui K, Hiratsu K, Shinshi H, Ohme-Takagi M: Repression domains of class II ERF transcriptional repressors share an essential motif for active repression. *Plant Cell* 2001, 13:1959-1968.

SUPPLEMENTARY INFORMATION

Nucleotide Sequences:

44246 dCas9

TGAAAGACCCCACCTGTAGGTTTGGCAAGCTAGCTTAAGTAACGCCATTTTGCAAGG

CATGGAAAATACATAACTGAGAATAGAGAAGTTCAGATCAAGGTTAGGAACAGAGAG

ACAGCAGAATATGGGCCAAACAGGATATCTGTGGTAAGCAGTTCCTGCCCCGGCTCA

GGGCCAAGAACAGATGGTCCCCAGATGCGGTCCCGCCCTCAGCAGTTTCTAGAGAAC

CATCAGATGTTTCCAGGGTGCCCCAAGGACCTGAAATGACCCTGTGCCTTATTTGAA

CTAACCAATCAGTTCGCTTCTCGCTTCTGTTCGCGCGCTTCTGCTCCCCGAGCTCAA

TAAAAGAGCCCACAACCCCTCACTCGGCGCGCCAGTCCTCCGATAGACTGCGTCGCC

CGGGTACCCGTATTCCCAATAAAGCCTCTTGCTGTTTGCATCCGAATCGTGGACTCG

CTGATCCTTGGGAGGGTCTCCTCAGATTGATTGACTGCCCACCTCGGGGGTCTTTCA

TTTGGAGGTTCCACCGAGATTTGGAGACCCCTGCCCAGGGACCACCGACCCCCCCGC

CGGGAGGTAAGCTGGCCAGCGGTCGTTTCGTGTCTGTCTCTGTCTTTGTGCGTGTTT

GTGCCGGCATCTAATGTTTGCGCCTGCGTCTGTACTAGTTAGCTAACTAGCTCTGTA

TCTGGCGGACCCGTGGTGGAACTGACGAGTTCTGAACACCCGGCCGCAACCCTGGGA

GACGTCCCAGGGACTTTGGGGGCCGTTTTTGTGGCCCGACCTGAGGAAGGGAGTCGA

TGTGGAATCCGACCCCGTCAGGATATGTGGTTCTGGTAGGAGACGAGAACCTAAAAC

AGTTCCCGCCTCCGTCTGAATTTTTGCTTTCGGTTTGGAACCGAAGCCGCGCGTCTT

GTCTGCTGCAGCGCTGCAGCATCGTTCTGTGTTGTCTCTGTCTGACTGTGTTTCTGT

ATTTGTCTGAAAATTAGGGCCAGACTGTTACCACTCCCTTAAGTTTGACCTTAGGTC

ACTGGAAAGATGTCGAGCGGATCGCTCACAACCAGTCGGTAGATGTCAAGAAGAGAC

GTTGGGTTACCTTCTGCTCTGCAGAATGGCCAACCTTTAACGTCGGATGGCCGCGAG

ACGGCACCTTTAACCGAGACCTCATCACCCAGGTTAAGATCAAGGTCTTTTCACCTG

GCCCGCATGGACACCCAGACCAGGTCCCCTACATCGTGACCTGGGAAGCCTTGGCTT

TTGACCCCCCTCCCTGGGTCAAGCCCTTTGTACACCCTAAGCCTCCGCCTCCTCTTC

CTCCATCCGCCCCGTCTCTCCCCCTTGAACCTCCTCGTTCGACCCCGCCTCGATCCT

CCCTTTATCCAGCCCTCACTCCTTCTCTAGGCGCCGGAATTAGATCTCGCCACCATG
GACAAGAAGTATTCTATCGGACTGGCCATCGGGACTAATAGCGTCGGGTGGGCCGTG
ATCACTGACGAGTACAAGGTGCCCTCTAAGAAGTTCAAGGTGCTCGGGAACACCGAC
CGGCATTCCATCAAGAAAAATCTGATCGGAGCTCTCCTCTTTGATTCAGGGGAGACC
GCTGAAGCAACCCGCCTCAAGCGGACTGCTAGACGGCGGTACACCAGGAGGAAGAAC
CGGATTTGTTACCTTCAAGAGATATTCTCCAACGAAATGGCAAAGGTCGACGACAGC
TTCTTCCATAGGCTGGAAGAATCATTCCTCGTGGAAGAGGATAAGAAGCATGAACGG
CATCCCATCTTCGGTAATATCGTCGACGAGGTGGCCTATCACGAGAAATACCCAACC
ATCTACCATCTTCGCAAAAAGCTGGTGGACTCAACCGACAAGGCAGACCTCCGGCTT
ATCTACCTGGCCCTGGCCCACATGATCAAGTTCAGAGGCCACTTCCTGATCGAGGGC
GACCTCAATCCTGACAATAGCGATGTGGATAAACTGTTCATCCAGCTGGTGCAGACT
TACAACCAGCTCTTTGAAGAGAACCCCATCAATGCAAGCGGAGTCGATGCCAAGGCC
ATTCTGTCAGCCCGGCTGTCAAAGAGCCGCAGACTTGAGAATCTTATCGCTCAGCTG
CCGGGTGAAAAGAAAAATGGACTGTTCGGGAACCTGATTGCTCTTTCACTTGGGCTG
ACTCCCAATTTCAAGTCTAATTTCGACCTGGCAGAGGATGCCAAGCTGCAACTGTCC
AAGGACACCTATGATGACGATCTCGACAACCTCCTGGCCCAGATCGGTGACCAATAC
GCCGACCTTTTCCTTGCTGCTAAGAATCTTTCTGACGCCATCCTGCTGTCTGACATT
CTCCGCGTGAACACTGAAATCACCAAGGCCCCTCTTTCAGCTTCAATGATTAAGCGG
TATGATGAGCACCACCAGGACCTGACCCTGCTTAAGGCACTCGTCCGGCAGCAGCTT
CCGGAGAAGTACAAGGAAATCTTCTTTGACCAGTCAAAGAATGGATACGCCGGCTAC
ATCGACGGAGGTGCCTCCCAAGAGGAATTTTATAAGTTTATCAAACCTATCCTTGAG
AAGATGGACGGCACCGAAGAGCTCCTCGTGAAACTGAATCGGGAGGATCTGCTGCGG
AAGCAGCGCACTTTCGACAATGGGAGCATTCCCCACCAGATCCATCTTGGGGAGCTT
CACGCCATCCTTCGGCGCCAAGAGGACTTCTACCCCTTTCTTAAGGACAACAGGGAG
AAGATTGAGAAAATTCTCACTTTCCGCATCCCCTACTACGTGGGACCCCTCGCCAGA
GGAAATAGCCGGTTTGCTTGGATGACCAGAAAGTCAGAAGAAACTATCACTCCCTGG
AACTTCGAAGAGGTGGTGGACAAGGGAGCCAGCGCTCAGTCATTCATCGAACGGATG
ACTAACTTCGATAAGAACCTCCCCAATGAGAAGGTCCTGCCGAAACATTCCCTGCTC

TACGAGTACTTTACCGTGTACAACGAGCTGACCAAGGTGAAATATGTCACCGAAGGG

ATGAGGAAGCCCGCATTCCTGTCAGGCGAACAAAAGAAGGCAATTGTGGACCTTCTG

TTCAAGACCAATAGAAAGGTGACCGTGAAGCAGCTGAAGGAGGACTATTTCAAGAAA

ATTGAATGCTTCGACTCTGTGGAGATTAGCGGGGTCGAAGATCGGTTCAACGCAAGC

CTGGGTACCTACCATGATCTGCTTAAGATCATCAAGGACAAGGATTTTCTGGACAAT

GAGGAGAACGAGGACATCCTTGAGGACATTGTCCTGACTCTCACTCTGTTCGAGGAC

CGGGAAATGATCGAGGAGAGGCTTAAGACCTACGCCCATCTGTTCGACGATAAAGTG

ATGAAGCAACTTAAACGGAGAAGATATACCGGATGGGGACGCCTTAGCCGCAAACTC

ATCAACGGAATCCGGGACAAACAGAGCGGAAAGACCATTCTTGATTTCCTTAAGAGC

GACGGATTCGCTAATCGCAACTTCATGCAACTTATCCATGATGATTCCCTGACCTTT

AAGGAGGACATCCAGAAGGCCCAAGTGTCTGGACAAGGTGACTCACTGCACGAGCAT

ATCGCAAATCTGGCTGGTTCACCCGCTATTAAGAAGGGTATTCTCCAGACCGTGAAA

GTCGTGGACGAGCTGGTCAAGGTGATGGGTCGCCATAAACCAGAGAACATTGTCATC

GAGATGGCCAGGGAAAACCAGACTACCCAGAAGGGACAGAAGAACAGCAGGGAGCGG

ATGAAAAGAATTGAGGAAGGGATTAAGGAGCTCGGGTCACAGATCCTTAAAGAGCAC

CCGGTGGAAAACACCCAGCTTCAGAATGAGAAGCTCTATCTGTACTACCTTCAAAAT

GGACGCGATATGTATGTGGACCAAGAGCTTGATATCAACAGGCTCTCAGACTACGAC

GTGGACGCCATCGTCCCTCAGAGCTTCCTCAAAGACGACTCAATTGACAATAAGGTG

CTGACTCGCTCAGACAAGAACCGGGGAAAGTCAGATAACGTGCCCTCAGAGGAAGTC

GTGAAAAAGATGAAGAACTATTGGCGCCAGCTTCTGAACGCAAAGCTGATCACTCAG

CGGAAGTTCGACAATCTCACTAAGGCTGAGAGGGGCGGACTGAGCGAACTGGACAAA

GCAGGATTCATTAAACGGCAACTTGTGGAGACTCGGCAGATTACTAAACATGTCGCC

CAAATCCTTGACTCACGCATGAATACCAAGTACGACGAAAACGACAAACTTATCCGC

GAGGTGAAGGTGATTACCCTGAAGTCCAAGCTGGTCAGCGATTTCAGAAAGGACTTT

CAATTCTACAAAGTGCGGGAGATCAATAACTATCATCATGCTCATGACGCATATCTG

AATGCCGTGGTGGGAACCGCCCTGATCAAGAAGTACCCAAAGCTGGAAAGCGAGTTC

GTGTACGGAGACTACAAGGTCTACGACGTGCGCAAGATGATTGCCAAATCTGAGCAG

GAGATCGGAAAGGCCACCGCAAAGTACTTCTTCTACAGCAACATCATGAATTTCTTC

AAGACCGAAATCACCCTTGCAAACGGTGAGATCCGGAAGAGGCCGCTCATCGAGACT
AATGGGGAGACTGGCGAAATCGTGTGGGACAAGGGCAGAGATTTCGCTACCGTGCGC
AAAGTGCTTTCTATGCCTCAAGTGAACATCGTGAAGAAAACCGAGGTGCAAACCGGA
GGCTTTTCTAAGGAATCAATCCTCCCCAAGCGCAACTCCGACAAGCTCATTGCAAGG
AAGAAGGATTGGGACCCTAAGAAGTACGGCGGATTCGATTCACCAACTGTGGCTTAT
TCTGTCCTGGTCGTGGCTAAGGTGGAAAAAGGAAAGTCTAAGAAGCTCAAGAGCGTG
AAGGAACTGCTGGGTATCACCATTATGGAGCGCAGCTCCTTCGAGAAGAACCCAATT
GACTTTCTCGAAGCCAAAGGTTACAAGGAAGTCAAGAAGGACCTTATCATCAAGCTC
CCAAAGTATAGCCTGTTCGAACTGGAGAATGGGCGGAAGCGGATGCTCGCCTCCGCT
GGCGAACTTCAGAAGGGTAATGAGCTGGCTCTCCCCTCCAAGTACGTGAATTTCCTC
TACCTTGCAAGCCATTACGAGAAGCTGAAGGGGAGCCCCGAGGACAACGAGCAAAAG
CAACTGTTTGTGGAGCAGCATAAGCATTATCTGGACGAGATCATTGAGCAGATTTCC
GAGTTTTCTAAACGCGTCATTCTCGCTGATGCCAACCTCGATAAAGTCCTTAGCGCA
TACAATAAGCACAGAGACAAACCAATTCGGGAGCAGGCTGAGAATATCATCCACCTG
TTCACCCTCACCAATCTTGGTGCCCCTGCCGCATTCAAGTACTTCGACACCACCATC
GACCGGAAACGCTATACCTCCACCAAAGAAGTGCTGGACGCCACCCTCATCCACCAG
AGCATCACCGGACTTTACGAAACTCGGATTGACCTCTCACAGCTCGGAGGGGATGAG
GGAGCTGATCCAAAAAAGAAGAGAAAGGTAGATCCAAAAAAGAAGAGAAAGGTAGAT
CCAAAAAAGAAGAGAAAGGTATAGAATTCTACCGGGTAGGGGAGGCGCTTTTCCCAA
GGCAGTCTGGAGCATGCGCTTTAGCAGCCCCGCTGGGCACTTGGCGCTACACAAGTG
GCCTCTGGCCTCGCACACATTCCACATCCACCGGTAGGCGCCAACCGGCTCCGTTCT
TTGGTGGCCCCTTCGCGCCACCTTCTACTCCTCCCCTAGTCAGGAAGTTCCCCCCCG
CCCCGCAGCTCGCGTCGTGCAGGACGTGACAAATGGAAGTAGCACGTCTCACTAGTC
TCGTGCAGATGGACAGCACCGCTGAGCAATGGAAGCGGGTAGGCCTTTGGGGCAGCG
GCCAATAGCAGCTTTGCTCCTTCGCTTTCTGGGCTCAGAGGCTGGGAAGGGGTGGGT
CCGGGGGCGGGCTCAGGGGCGGGCTCAGGGGCGGGGCGGGCGCCCGAAGGTCCTCCG
GAGGCCCGGCATTCTGCACGCTTCAAAAGCGCACGTCTGCCGCGCTGTTCTCCTCTT
CCTCATCTCCGGGCCTTTCGACCTGCAGCCCAAGCTTACCATGACCGAGTACAAGCC

CACGGTGCGCCTCGCCACCCGCGACGACGTCCCCAGGGCCGTACGCACCCTCGCCGC

CGCGTTCGCCGACTACCCCGCCACGCGCCACACCGTCGATCCGGACCGCCACATCGA

GCGGGTCACCGAGCTGCAAGAACTCTTCCTCACGCGCGTCGGGCTCGACATCGGCAA

GGTGTGGGTCGCGGACGACGGCGCCGCGGTGGCGGTCTGGACCACGCCGGAGAGCGT

CGAAGCGGGGGCGGTGTTCGCCGAGATCGGCCCGCGCATGGCCGAGTTGAGCGGTTC

CCGGCTGGCCGCGCAGCAACAGATGGAAGGCCTCCTGGCGCCGCACCGGCCCAAGGA

GCCCGCGTGGTTCCTGGCCACCGTCGGCGTCTCGCCCGACCACCAGGGCAAGGGTCT

GGGCAGCGCCGTCGTGCTCCCCGGAGTGGAGGCGGCCGAGCGCGCCGGGGTGCCCGC

CTTCCTGGAGACCTCCGCGCCCCGCAACCTCCCCTTCTACGAGCGGCTCGGCTTCAC

CGTCACCGCCGACGTCGAGGTGCCCGAAGGACCGCGCACCTGGTGCATGACCCGCAA

GCCCGGTGCCTGACGCCCGCCCCACGACCCGCAGCGCCCGACCGAAAGGAGCGCACG

ACCCCATGCATCGATAAAATAAAAGATTTTATTTAGTCTCCAGAAAAAGGGGGGAAT

GAAAGACCCCACCTGTAGGTTTGGCAAGCTAGCTTAAGTAACGCCATTTTGCAAGGC

ATGGAAAATACATAACTGAGAATAGAGAAGTTCAGATCAAGGTTAGGAACAGAGAGA

CAGCAGAATATGGGCCAAACAGGATATCTGTGGTAAGCAGTTCCTGCCCCGGCTCAG

GGCCAAGAACAGATGGTCCCCAGATGCGGTCCCGCCCTCAGCAGTTTCTAGAGAACC

ATCAGATGTTTCCAGGGTGCCCCAAGGACCTGAAATGACCCTGTGCCTTATTTGAAC

TAACCAATCAGTTCGCTTCTCGCTTCTGTTCGCGCGCTTCTGCTCCCCGAGCTCAAT

AAAAGAGCCCACAACCCCTCACTCGGCGCGCCAGTCCTCCGATAGACTGCGTCGCCC

GGGTACCCGTGTATCCAATAAACCCTCTTGCAGTTGCATCCGACTTGTGGTCTCGCT

GTTCCTTGGGAGGGTCTCCTCTGAGTGATTGACTACCCGTCAGCGGGGGTCTTTCAT

GGGTAACAGTTTCTTGAAGTTGGAGAACAACATTCTGAGGGTAGGAGTCGAATATTA

AGTAATCCTGACTCAATTAGCCACTGTTTTGAATCCACATACTCCAATACTCCTGAA

ATAGTTCATTATGGACAGCGCAGAAGAGCTGGGGAGAATTAATTCGTAATCATGGTC

ATAGCTGTTTCCTGTGTGAAATTGTTATCCGCTCACAATTCCACACAACATACGAGC

CGGAAGCATAAAGTGTAAAGCCTGGGGTGCCTAATGAGTGAGCTAACTCACATTAAT

TGCGTTGCGCTCACTGCCCGCTTTCCAGTCGGGAAACCTGTCGTGCCAGCTGCATTA

ATGAATCGGCCAACGCGCGGGGAGAGGCGGTTTGCGTATTGGGCGCTCTTCCGCTTC

CTCGCTCACTGACTCGCTGCGCTCGGTCGTTCGGCTGCGGCGAGCGGTATCAGCTCA
CTCAAAGGCGGTAATACGGTTATCCACAGAATCAGGGGATAACGCAGGAAAGAACAT
GTGAGCAAAAGGCCAGCAAAAGGCCAGGAACCGTAAAAAGGCCGCGTTGCTGGCGTT
TTTCCATAGGCTCCGCCCCCCTGACGAGCATCACAAAAATCGACGCTCAAGTCAGAG
GTGGCGAAACCCGACAGGACTATAAAGATACCAGGCGTTTCCCCCTGGAAGCTCCCT
CGTGCGCTCTCCTGTTCCGACCCTGCCGCTTACCGGATACCTGTCCGCCTTTCTCCC
TTCGGGAAGCGTGGCGCTTTCTCATAGCTCACGCTGTAGGTATCTCAGTTCGGTGTA
GGTCGTTCGCTCCAAGCTGGGCTGTGTGCACGAACCCCCCGTTCAGCCCGACCGCTG
CGCCTTATCCGGTAACTATCGTCTTGAGTCCAACCCGGTAAGACACGACTTATCGCC
ACTGGCAGCAGCCACTGGTAACAGGATTAGCAGAGCGAGGTATGTAGGCGGTGCTAC
AGAGTTCTTGAAGTGGTGGCCTAACTACGGCTACACTAGAAGGACAGTATTTGGTAT
CTGCGCTCTGCTGAAGCCAGTTACCTTCGGAAAAAGAGTTGGTAGCTCTTGATCCGG
CAAACAAACCACCGCTGGTAGCGGTGGTTTTTTTGTTTGCAAGCAGCAGATTACGCG
CAGAAAAAAAGGATCTCAAGAAGATCCTTTGATCTTTTCTACGGGGTCTGACGCTCA
GTGGAACGAAAACTCACGTTAAGGGATTTTGGTCATGAGATTATCAAAAAGGATCTT
CACCTAGATCCTTTTAAATTAAAAATGAAGTTTTAAATCAATCTAAAGTATATATGA
GTAAACTTGGTCTGACAGTTACCAATGCTTAATCAGTGAGGCACCTATCTCAGCGAT
CTGTCTATTTCGTTCATCCATAGTTGCCTGACTCCCCGTCGTGTAGATAACTACGAT
ACGGGAGGGCTTACCATCTGGCCCCAGTGCTGCAATGATACCGCGAGACCCACGCTC
ACCGGCTCCAGATTTATCAGCAATAAACCAGCCAGCCGGAAGGGCCGAGCGCAGAAG
TGGTCCTGCAACTTTATCCGCCTCCATCCAGTCTATTAATTGTTGCCGGGAAGCTAG
AGTAAGTAGTTCGCCAGTTAATAGTTTGCGCAACGTTGTTGCCATTGCTACAGGCAT
CGTGGTGTCACGCTCGTCGTTTGGTATGGCTTCATTCAGCTCCGGTTCCCAACGATC
AAGGCGAGTTACATGATCCCCCATGTTGTGCAAAAAAGCGGTTAGCTCCTTCGGTCC
TCCGATCGTTGTCAGAAGTAAGTTGGCCGCAGTGTTATCACTCATGGTTATGGCAGC
ACTGCATAATTCTCTTACTGTCATGCCATCCGTAAGATGCTTTTCTGTGACTGGTGA
GTACTCAACCAAGTCATTCTGAGAATAGTGTATGCGGCGACCGAGTTGCTCTTGCCC
GGCGTCAATACGGGATAATACCGCGCCACATAGCAGAACTTTAAAAGTGCTCATCAT

TGGAAAACGTTCTTCGGGGCGAAAACTCTCAAGGATCTTACCGCTGTTGAGATCCAG

TTCGATGTAACCCACTCGTGCACCCAACTGATCTTCAGCATCTTTTACTTTCACCAG

CGTTTCTGGGTGAGCAAAAACAGGAAGGCAAAATGCCGCAAAAAAGGGAATAAGGGC

GACACGGAAATGTTGAATACTCATACTCTTCCTTTTTCAATATTATTGAAGCATTTA

TCAGGGTTATTGTCTCATGAGCGGATACATATTTGAATGTATTTAGAAAAATAAACA

AATAGGGGTTCCGCGCACATTTCCCCGAAAAGTGCCACCTGACGTCTAAGAAACCAT

TATTATCATGACATTAACCTATAAAAATAGGCGTATCACGAGGCCCTTTCGTCTCGC

GCGTTTCGGTGATGACGGTGAAAACCTCTGACACATGCAGCTCCCGGAGACGGTCAC

AGCTTGTCTGTAAGCGGATGCCGGGAGCAGACAAGCCCGTCAGGGCGCGTCAGCGGG

TGTTGGCGGGTGTCGGGGCTGGCTTAACTATGCGGCATCAGAGCAGATTGTACTGAG

AGTGCACCATATGCGGTGTGAAATACCGCACAGATGCGTAAGGAGAAAATACCGCAT

CAGGCGCCATTCGCCATTCAGGCTGCGCAACTGTTGGGAAGGGCGATCGGTGCGGGC

CTCTTCGCTATTACGCCAGCTGGCGAAAGGGGGATGTGCTGCAAGGCGATTAAGTTG

GGTAACGCCAGGGTTTTCCCAGTCACGACGTTGTAAAACGACGGCGCAAGGAATGGT

GCATGCAAGGAGATGGCGCCCAACAGTCCCCCGGCCACGGGGCCTGCCACCATACCC

ACGCCGAAACAAGCGCTCATGAGCCCGAAGTGGCGAGCCCGATCTTCCCCATCGGTG

ATGTCGGCGATATAGGCGCCAGCAACCGCACCTGTGGCGCCGGTGATGCCGGCCACG

ATGCGTCCGGCGTAGAGGCGATTAGTCCAATTTGTTAAAGACAGGATATCAGTGGTC

CAGGCTCTAGTTTTGACTCAACAATATCACCAGCTGAAGCCTATAGAGTACGAGCCA

TAGATAAAATAAAAGATTTTATTTAGTCTCCAGAAAAGGGGGGAA

N. benthamiana PDS gene with promoter sequence

```
TAACGAAAACATGGAATATGCGGGTAATAACCAAACAAGGCACAATAATACTAATAA
GGTTTGATATATGGCATATTTAGGTAAAAATCCCTTTTAATAAAAGTATCAGGACGA
GAAACAAAAAAGAAGATTACTACTACTAGAAAACCACGGGTCAGGTTGGATTACGT
GGATCCTCTATGACCCAGTAACCCATGTGGGAGATGGGAGCAAAGTGGTCAAACTTT
AGAAGGAATGAGCAAAGCAAGAAATTAAAAAGAGAGAGAGGTGCTTTATCCATCAAA
TGTGGCTATGGTAGGAAGAGCCAATGGTGGGACATTTTTGGAGTGTAGCCAAAACAT
AAAGGAAGGTCCAGTGCGAGTTACTGCAAATTGAGTTGGGAGTGAGGATTAAAGAAG
ATAGTAACATATTTCTAGCTAAATAGCAAACAAATGATCCGTTAACAGAAGTGGCCA
AACCACCAAATTCAGGCATCTCCACCAAATATTAGTTTTTTATACACAAAGATTCA
ACACAAACAGTTAAGTACTTCTTTAATCGTTCCTAATTCTTTGTTCAGGGGTATCTT
TTTGTGGGTAACGGCCAAACCACCACAAATTTTCAGTTCCCACTCTTAACTCTTTCA
ACTTCAACACAACAAATTAGTATTTGCTTTTCCTTCTTTGCTTATCTAGTGCATAAC
GATTTTCTACAACTTTAGCATAGTCCACAACGTGAAACACAACTCCTTGGCGGTTTA
TACCGAGGTAAGAAATGATTTTGGTTTCTTTGGTTACATCAGCTGAATGCTTTGCTT
GAGAAAAGCTCTCTTTTTCCCGTTTAGGATCTTGTTTATTTGCTTTCGTTTTTCTAC
TCGTTTGAATTTTAACTTGATTTTGTGGGTGAAGGCTAATTTTTCTCATAGTGTAAG
AACAAGTTTCATATGTACTGTAAAAGCTAGAATCTTTTTTACTTTTGCATATAAATT
TGTGTAATAAATGCTTAAGAACCAGAATATTTGAAAAAGATAAGGAATTTTGCATAG
TATTTAGGTTCACAAGTGGGACAATCTTCTTACACTGAAATATCTTTATGTCAGGCT
TAATTTACTGCTATCTTGTTCAATAAAATGCCCCAAATTGGACTTGTTTCTGCCGTT
AATTTGAGAGTCCAAGGTAATTCAGCTTATCTTTGGAGCTCGAGGTCTTCGTTGGGA
ACTGAAAGTCAAGATGTTTGCTTGCAAAGGAATTTGTTATGTTTTGGTAGTAGCGAC
TCCATGGGGCATAAGTTAAGGATTCGTACTCCAAGTGCCACGACCCGAAGATTGACA
AAGGACTTTAATCCTTTAAAGGTTTGTTTTGAATGCGAAAGTGTGATGCTGGATTTA
TGATCGTGGGCATATATCCTCTAAAATAAGAGATGTATATCTTGCCATTCAGGTAGT
CTGCATTGATTATCCAAGACCAGAGCTAGACAATACAGTTAACTATTTGGAGGCGGC
GTTATTATCATCATCGTTTCGTACTTCCTCACGCCCAACTAAACCATTGGAGATTGT
```

TATTGCTGGTGCAGGTGATTTTTTCCAGCCATCTATATTTGTAGTTTTCATTTTTCT

TTCTTTGGAAGGAAGATCATTCTATTAGTTATATTATCACTAGAATATTTACCTGTA

CATTCTTTTCTGATTAACTGTTTTGGACCGCAAAATTTTAGGTTCTTACTTCTTCGC

CATTTTGCAACTAATCAGCAATTAGGAGCGGTTTGAAAACTAGTTTGTTTTGAACTA

TTTTTGCCGTCACTCTATTTATATACTGTTGAATTGTCCCAAATCGGTGGAATTTGA

GGTCCTTGGTCTCATCTCATAAGCTAGCTTTTGGGGTTGAGTTACCACATCGGTGGG

ATTTGAGGTCCTTCGTCTCCTTATATGTTCTTGGACAAGCTTCACCTCATAAGCTAG

CTTTTGGGGTTAGAGTTAGGCCCAAGGTCCATTTATCATATGCTTGTCTATTCTCTC

TTATCATCTGAGCCATGATAAGCGGGTGAACGTGCTGTCTATTGGGTGGCATGTCCA

AAGGATCATTCTGAAATATTGGAGGCAAATGAACCAATACCTTGTGCAAGATTGATC

TCACTATACCTATAATCAGAGTACTGAGTTCCAAAAATTTCAAAACCCATTGAAAAG

TCAAACGAGTTACATATAGGGGTTGCACTCTTCTACGGCTTGCAATATGTGAGAAAA

AGATGAGAAGTCGATCTTCATATTTCATCTTTACTAGGCTGGACCATTGACTGGTTA

GCAGTTTTGAACTTGTTCTTCAACTTGGCTTGCATGGTACTGTGCCGATCATTTCTT

TTGTATTGTCATCAGCTGGTTGATTATCTGAGTACCTAAAGAAAGAATGTTATATGC

ATGATATATTCTACTGTACTATAAAGATATAATAAAGAATGCTAGCCGAGGTACTG

CATGGCCTTTTCAGATAAATAGAAGCTGTAGCATGATTCTAATTCAATTTTTTTGGG

AATATCAGGTTTGGGTGGTTTGTCTACAGCAAAATATCTGGCAGATGCTGGTCACAA

ACCGATATTGCTGGAGGCAAGAGATGTCCTAGGTGGGAAGGTGAAGAATATCCAATC

TTTCCTTTAATTTTATTCCTTTTTCTTTTGTGTCCTTCCCTATTGATAGTCCCTTTT

CAGGAAGGCTTCTGTTTGTTTTATTTGAAATCATTTTTCATACTCTTTAAGCATTCA

GTTGCTCAAACAATTGCAAGGATATTCACTATTCCTAATTTTGACCGTCTTCTTTTC

TCTCAGTTTAGTTTTATTCCCCTCTCTTTTTGAAGGAAATAGATCTGTCCTAAAAAT

TTCCAGCTTTACTACTAATAGTGTTAATTGTCGATAAAATAGTACATCATATTAGGT

AAAAGATATGGACTGTATATTATTATCATTCTCTATTATTTTAAACTGAGTCAATTT

TAACCGTCCTGTTGGGTGCATTTCTCATATAAACAGTCTTTTCTGTGAGATGCTATG

TGAATTAGCTGATTGTTTTGGTATAGAGCACTATGTTAGTCAGTTTTATCTTACTGA

AGCAGTCACCAAGAGTCTAGTTGTATAGGCTAGAAGATTGAATTAGCATTAATCTTT

ATGTGTTCTGCACCTGAATACTTGTACCTCCCTTTTAGGTAGCTGCATGGAAAGATG
ATGATGGAGATTGGTACGAGACTGGGTTGCACATATTCTGTAAGTTTGACTCCTCAA
GAATGCTACTTTAATCTTCTAATACAGTCATAGCAATTTCTTTCAAGATCTCTTTTA
TTAATCAGATAGCTATCCCTGTTTGTCTTTTGTCTTTTGCAAATAGCCAATTTTTGT
CAGTCGATCTGTATTCTGCCTTGCCTCTCTTTATTTATCTGCTAACTCGTATGGTGA
CTCATACAAGTTGGTGCATCTCCTTTAAGTTGGGGCTTACCCAAATATGCAGAACCT
GTTTGGAGAACTAGGGATTGATGATCGGTTGCAGTGGAAGGAACATTCAATGATATT
TGCGATGCCTAACAAGCCAGGGGAGTTCAGCCGCTTTGATTTTCCTGAAGCTCTTCC
TGCGCCATTAAATGGTAAGTACTTAATCATGAGTAAATTTCTCCCTTCAGCGTTGAT
TATGCAAACTTCCCCAATAAGGTATGAAATTGATTAGTCTTAATACCCTGGCACATT
GCTAACATCAAAAGAACATAAAGGTTCATTACGTCTTGATCAGAATTTCTGCATGTA
GCTAAAGTGAATGAGTGTCTGTATAGATTTTTACACATTGCAAGCATAAGCCTGTTA
TGTTATCTCTTTTTTTCATTTCTCTACCTGTATCTCTTATTCTCATTTCTCTATCTA
TGCGTTATTACTTCTACAGGAATTTTGGCCATACTAAAGAACAACGAAATGCTTACG
TGGCCCGAGAAAGTCAAATTTGCTATTGGACTCTTGCCAGCAATGCTTGGAGGGCAA
TCTTATGTTGAAGCTCAAGACGGTTTAAGTGTTAAGGACTGGATGAGAAAGCAAGTA
TGTGATCGTTTTATCTTATTCTTTAAAGTTCATAACCTTGAGGACATAGTTGACTTG
CATATTGTTGATTTAACATGTTCGAATTGTCTACCTGCCTTTCTTTTTCTAACAACA
TAGATCTTACAATCTCAGCAGCAGCTATTTGCTTAATGCTTTTCAGGGTGTGCCTGA
TAGGGTGACAGATGAGGTGTTCATTGCCATGTCAAAGGCACTTAACTTCATAAACCC
TGACGAGCTTTCGATGCAGTGCATTTTGATTGCTTTGAACAGATTTCTTCAGGTTAG
AATCCTGATCCACCCTCAAAACAAAAAGAGAGAAAGGGATATAATCCTACCAAAGCT
GTAAATCATGTTAGGGACCTGACATATCGGTGCAGGAAACTTATGAGTGAACTTGTC
CACTCTGTTTAACTTTTCTGATATATTTGAATTATTAATCTGCAGGAGAAACATGGT
TCAAAAATGGCCTTTTTAGATGGTAACCCTCCTGAGAGACTTTGCATGCCGATTGTG
GAACATATTGAGTCAAAAGGTGGCCAAGTCAGACTAAACTCACGAATAAAAAAGATC
GAGCTGAATGAGGATGGAAGTGTCAAATGTTTTATACTGAATAATGGCAGTACAATT
AAAGGAGATGCTTTTGTGTTTGCCACTCCAGGTATAATATCCATTATACTAGTATCG

ATGCTTCCAGTTTTCACATTTTTAGTATGAGTACAATTAAAGGAGATGCTTTTGTGT

TTGCCACTCCAGGTATAATATCCATTACACTAGTATGACGCTTCCAGTTTTCACATT

TTAATATGAATTTATAGTTTTTTGCTGACTTTTGATTATCCAATTAGTGGATATCTT

GAAGCTTCTTTTGCCTGAAGACTGGAAAGAGATCCCATATTTCCAAAAGTTGGAGAA

GCTAGTGGGAGTTCCTGTGATAAATGTCCATATATGGTTAGTGATGAAAATTTTGCT

TTTCAGTGTTTGGTCTTCCTCTAGCATATCTATGTATGTGCATGTTAATGTCTATAC

GTACATGTTTATGTGGTCCTCCCGTATTGTGTTTACTTCCCTTGAATGAGGAACTTA

TGGATGTACGCTTTTCCAACTTTGATTGTACACATTGCAATTGTCTGTTCAACTTTG

AGGAGCAGAACTTCCATTGTTTAGCTATTAGTGGCTGAGATTCCTGCTGAAAAGATT

TGTATAAATTTAATTTGCAGGTTTGACAGAAAACTGAAGAACACATCTGATAATCTG

CTCTTCAGCAGGTTCATTTTTGATCAATTTTATTGTTCCAGACCAGTTTCTGCGTGT

CCATGACTACATTCTCATATTAGCTCCCCCCCCCCCCCCCCCCCCCCCCCNNNNNNNN

NNCCCCCGGTCTC

TTTTTTGCCATTTAAATGAGACCTTACAATTTGTTTAGTACTCTACCATAGTTTTTT

AATCAATAAGCCAAAGGGGAAAAACTAATAAAAGTGTATAAAATTTCTTCCTGTATT

AGTCCAATTCTTTCGCAACTTATATTGTTAATTATTATTTATCTTTTGGATTGAAAT

GGATTTTGTATATCTAATAATATAAACAAATATATCTCTTCCTCTTATAAGATTTTT

CACCATAGAAAAATGCTCCCATAAGGTCAGTCATTCTGGCTAAATATCCCACACTTC

AACCATTGAGATATTTTGTTCTTTGCATCCAGGAATACATTTGGCATCAATAGATAG

GAATCAATGAAGATATATTATCAATTTCCTGCAAGTTTCTTGGCACTAGAAACATTA

GATCCATATCATGTAAATTGCCTTTGTTAAATTGAAGGTCTATGAAATTTGGGTTGG

TTTGAAAACCTTTTGTTTTTCCCCCCCACATCCCTAATCGTTTATTTAGTCAAGGTC

AGACCTGACATGTTATGATGACCATTTCTCCAAGGCATTTATAATGGACTGGAGTAT

CCATGCCACATTTCATCAGCTACATGTCGATTATGTTCCCCTACTTTTAAATGGCAC

CATTGTTGGTGGAGCAAGATTATAGATTTTCCTGATACTTGTATGGGTTCCCTTGCT

CAATCTCTCTTTTACTTCATGCAGAAGCCCGTTGCTCAGTGTGTACGCTGACATGTC

TGTTACATGTAAGGTATTGACTCGTCTGTACCATTATACTGGTCTAATCTGTTGGGT

ATGAGTTGCTGGTAAATTGCATAATGCTTGTTGGATTTGTGTGTGAGTTGCTGCTAG

ATCTATGTCCTGCTATATTTATGTATGAGTTGCTGTTGTTGCAATCTTCATTTCGAA
TGCATAATGATATAGGTTCTGTATGTACGGAATAGTCAGGACAATGCTCCTGTCTGT
GCACGGGGGCTCTACAGGAAGCAACTTTCGGAGGAGAAGTACAGAAAGTGTGATGAA
TCTTAAGGCGGTTAAAGTAGTTTCTTTTAGCTAAATTTTGAAATAATTTGAAGGAGG
GGAAAACGCTCTCTCAGTCTGTGGTTGCATTGGTTGTGGGNNNNNNNNNNNNNNNNNN
NNNNNNNNNNNNNNNNNNNNNNNNNNNNNNNNNNNNNNNNNNNNNNNNNNNNNNNNNNN
NNNNNNNNNNNNNNNNNNNNNNNNNNNNNNNNNNNNNNNNNNNNNNNNNNNNNNNNNNN
NNNNNNNNNNNNNNNNNNNNNNNNNNNNNNNNNNNNNNNNNNNNNNNNNNNNNNNNNNN
NNNNNNNNNNNNNNNNNNNNNNNNNNNNNNNNNNNNNNNNNNNNNNNNNNNNNNNNNNN
NNNNNNNNNNNNNNNNNNNNNNNNNNNNNNNNNNNNNNNNNNNNNNNNNNNNNNNNNNN
NNNNNNNNNNNNNNNNNNNNNNNNNNNNNNNNNNNNNNNNNNNNNNNNNAGGGACCTTAAT
TCAGTGTTACCTGCATATAAATCAGACTAAAGCCTGGAGATCAGACGTTCTGCATAT
AAATAGATAATTAATAATGATCTCGTAATACTCTAAAGCCTGGAGATCAGACTGTTT
TAACTATCCTGAGATGATTACTTTTACTCTCGGATTAGCTTAGGCGAGCTGCAAGAC
TACATCGAATCTTTAGAAATGGGAACATAAAAAAGGTGCGAAGTGGGGAAGTGGCTG
AACAATAGGCATATGTGAGTGAGTGGGGAGTAAAATTACTTCCTTTACTTGGGTACA
GTCAAGAATGGATGACAGCTTAGCCCACTATATCTGTTCATGTGTTCTTTAGGGTCC
TCTGATATAACTGGTCTCTCTGCAGGAATATTACAACCCCAATCAGTCTATGTTGGA
ATTGGTATTTGCACCCGCAGAAGAGTGGATAAATCGTAGTGACTCAGAAATTATTGA
TGCTACAATGAAGGAACTAGCGAAGCTTTTCCCTGATGAAATTTCGGCAGATCAGAG
CAAAGCAAAAATATTGAAGTATCATGTTGTCAAAACCCCAAGGTCAGTAATCATTTT
GCTTTCATAGTTGTGTAGTATGCGAGAATTACTGTCCACGTGGAATCTATTCCTGTT
ATGAATCCTGATTAATCTGCTTTTTACTTTCAGGTCTGTTTATAAAACTGTGCCAGG
TTGTGAACCCTGTCGGCCCTTGCAAAGATCCCCTATAGAGGGTTTTTATTTAGCTGG
TGACTACACGAAACAGAAGTACTTGGCTTCAATGGAAGGTGCTGTCTTATCAGGAAA
GCTTTGTGCCGAAGCTATTGTACAGGTTAGCTCTCACATTTTTTTCCCTTCCATTGA
TAGTGTATTTGATTATATTTTGTCATCTTTGCTGCGGTAGAGAATTTTAGAAGCATT
TCTCAGACATTAGTTAGCAGAGTTACTCAGGATATCTGCAGTTTTGGAGCTTCAGTA

GTAGCATGATAAAATGCAGAGGATTGTGTTTTTTCATTCTTTATTAAACCTTGTGCC

AAAGGTCTTTTGGAAACAACCTCTCTACCCCGAGGTAGGGGTAAGGTCTGCGTACAT

ATTACCCTCCCCATACCCCATGCGTGGGATTATACTGGGTGGTTGTTGTATAAACCT

ATATCTCTATAATTTGCAGGATTACGAGTTACTTCTTGGCCGGAGCCAGAAGATGTT

GGCAGAAGCAAGCGTAGTTAGCATAGTGAACTAAAATGTTAATTCTGTACACAAAAT

TTAAGATGAAGGCGGCCACGCTGAATTAGCGTTGTACACAACTTATACAAGCACAGT

ACAACATTGAAACCAAATACGAGAAATGTTACACAAATATGTGCTGCTTTCCCTCCG

ATTTAGTTCACAAGTTACGGACTAATTATAAGATGGAATTGTATGCAAATTGATTCA

TTCAAAACCAAACTTTTAAGCGTCAGTTATACTAGCAATACCTATAAGACATTAAAC

CTTCACGTCTCAAGACATCAAAACTCGCTTTTAGAGTAAGTATCCAACTGAACCATG

CAAAACACAAAACTGAAACAACGAACCAAATGGAACTACTCAAATTTAAAAAGAAAG

AGAATTAAAATTTTAATTTCGTCAACTATTAAAATGTGCTTGCAAGTATGGGATTT

TTTTTTGAAATCCCCAAAAAACCTGTAGCCGCGACAACCTTAGCCTTTCGAGTGAGC

ACTTTGTGCTCATTGGATAAACCTCCCATTGTGTAATAGCTTGCAAACTACACACGG

GATATAAATTGTACTAGGCAAGCCCTGTGCGACAGGCTCGACCTAGAAGATATTGAG

GGAATCAATCCCAGGTTGTCTG

MluI-dCas9 – EDLL

TCTAAACGCGTCATTCTCGCTGATGCCAACCTCGATAAAGTCCTTAGCGCATACAAT
AAGCACAGAGACAAACCAATTCGGGAGCAGGCTGAGAATATCATCCACCTGTTCACC
CTCACCAATCTTGGTGCCCCTGCCGCATTCAAGTACTTCGACACCACCATCGACCGG
AAACGCTATACCTCCACCAAAGAAGTGCTGGACGCCACCCTCATCCACCAGAGCATC
ACCGGACTTTACGAAACTCGGATTGACCTCTCACAGCTCGGAGGGGATGAGGGAGCT
GATCCAAAAAAGAAGAGAAAGGTAGATCCAAAAAAGAAGAGAAAGGTAGATCCAAAA
AAGAAGAGAAAGGTAGGATCCGAAGTTTTTGAATTTGAATATCTTGATGATAAGGTT
CTTGAAGAACTTCTTGATTCTGAAGAAAGAAAGAGATGACTCGAGGAATTC

MluI-dCas9-TALE-AD-BamHI-XhoI

AGTTTTCTAAACGCGTCATTCTCGCTGATGCCAACCTCGATAAAGTCCTTAGCGCAT
ACAATAAGCACAGAGACAAACCAATTCGGGAGCAGGCTGAGAATATCATCCACCTGT
TCACCCTCACCAATCTTGGTGCCCCTGCCGCATTCAAGTACTTCGACACCACCATCG
ACCGGAAACGCTATACCTCCACCAAAGAAGTGCTGGACGCCACCCTCATCCACCAGA
GCATCACCGGACTTTACGAAACTCGGATTGACCTCTCACAGCTCGGAGGGGATGAGG
GAGCTGATCCAAAAAAGAAGAGAAAGGTAGATCCAAAAAAGAAGAGAAAGGTAGATC
CAAAAAAGAAGAGAAAGGTATCGATAGTCGCACAACTATCACGACCTGATCCCGCTC
TTGCAGCATTGACAAACGATCATTTAGTCGCACTTGCATGTTTAGGAGGACGACCAG
CACTTGATGCCGTTAAGAAAGGACTACCGCACGCCCCTGCATTGATTAAAAGAACAA
ACAGACGAATCCCGGAGAGAACTTCACATCGTGTAGCCGATCATGCTCAAGTCGTAA
GAGTTTTGGGTTTCTTCCAATGTCATTCCCACCCAGCTCAAGCTTTTGACGATGCAA
TGACTCAATTTGGAATGAGTAGACATGGACTCCTGCAATTATTTCGAAGGGTCGGAG
TTACAGAGCTCGAAGCCAGGTCAGGAACGCTGCCCCCCGCATCTCAACGATGGGATA
GAATTCTCCAAGCCTCTGGAATGAAAAGAGCTAAACCTTCACCAACGTCCACACAAA
CACCAGACCAAGCTTCTCTCCACGCTTTTGCCGACTCACTAGAGAGAGATCTAGATG
CACCGTCACCTATGCATGAAGGAGACCAAACAAGAGCCTCTTCAAGAAAACGTTCTC
GTTCTGATAGAGCTGTCACTGGACCTTCCGCCCAACAATCTTTCGAAGTCCGAGTTC

CTGAGCAACGAGATGCCCTACACCTGCCTTTGCTTTCTTGGGGAGTTAAGCGACCAC
GTACTAGAATTGGTGGACTACTCGATCCAGGTACACCAATGGATGCTGATCTCGTTG
CTTCCTCTACCGTAGTATGGGAGCAAGACGCAGACCCCTTCGCTGGAACTGCTGACG
ATTTCCCAGCCTTTAACGAGGAAGAATTGGCTTGGTTAATGGAACTTCTACCGCAAA
AGTAGGGATCCGGCTCGAGC

MluI-dCas9 – SRDX

TCTAAACGCGTCATTCTCGCTGATGCCAACCTCGATAAAGTCCTTAGCGCATACAAT
AAGCACAGAGACAAACCAATTCGGGAGCAGGCTGAGAATATCATCCACCTGTTCACC
CTCACCAATCTTGGTGCCCCTGCCGCATTCAAGTACTTCGACACCACCATCGACCGG
AAACGCTATACCTCCACCAAAGAAGTGCTGGACGCCACCCTCATCCACCAGAGCATC
ACCGGACTTTACGAAACTCGGATTGACCTCTCACAGCTCGGAGGGGATGAGGGAGCT
GATCCAAAAAGAAGAGAAAGGTAGATCCAAAAAGAAGAGAAAGGTAGATCCAAAA
AAGAAGAGAAAGGTAGGATCCCTGGATCTGGATCTGGAACTGCGCCTGGGCTTTGCG
TGATGACTCGAGGAATTC

Modified from li et.al 2013 (U6::gRNA)

GAGCTCGAATTCGGATCCAGAAATCTCAAAATTCCGGCAGAACAATTTTGAATCTCG
ATCCGTAGAAACGAGACGGTCATTGTTTTAGTTCCACCACGATTATATTTGAAATTT
ACGTGAGTGTGAGTGAGACTTGCATAAGAAAATAAAATCTTTAGTTGGGAAAAAATT
CAATAATATAAATGGGCTTGAGAAGGAAGCGAGGGATAGGCCTTTTTCTAAAATAGG
CCCATTTAAGCTATTAACAATCTTCAAAAGTACCACAGCGCTTAGGTAAAGAAAGCA
GCTGAGTTTATATATGGTTAGAGACGAAGTAGTGATT*GGGTCTTCGAGAAGACCT*GT
TTTAGAGCTAGAAATAGCAAGTTAAAATAAGGCTAGTCCGTTATCAACTTGAAAAAG
TGGCACCGAGTCGGTGCTTTTTTTAGACCCAGCTTTCTTGTACAAAGTTGGCATTAT
CTAGAAAGCTTGAGCTC

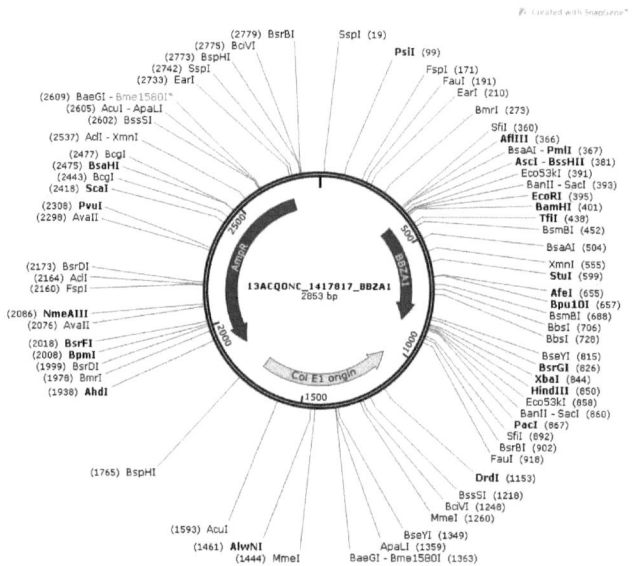

gRNA 1 (In 7)

GAGCTCGAATTCGGATCCAGAAATCTCAAAATTCCGGCAGAACAATTTTGAATCTCG
ATCCGTAGAAACGAGACGGTCATTGTTTTAGTTCCACCACGATTATATTTGAAATTT
ACGTGAGTGTGAGTGAGACTTGCATAAGAAAATAAAATCTTTAGTTGGGAAAAAATT
CAATAATATAAATGGGCTTGAGAAGGAAGCGAGGGATAGGCCTTTTTCTAAAATAGG
CCCATTTAAGCTATTAACAATCTTCAAAAGTACCACAGCGCTTAGGTAAAGAAAGCA
GCTGAGTTTATATATGGTTAGAGACGAAGTAGTGATTGACCGATAGATTAACCATTT
CGTTTTAGAGCTAGAAATAGCAAGTTAAAATAAGGCTAGTCCGTTATCAACTTGAAA
AAGTGGCACCGAGTCGGTGCTTTTTTTTAGACCCAGCTTTCTTGTACAAAGTTGGCAT
TATCTAGAAAGCTTGAGCTC

gRNA 2 (In 6)

GAGCTCGAATTCGGATCCAGAAATCTCAAAATTCCGGCAGAACAATTTTGAATCTCG
ATCCGTAGAAACGAGACGGTCATTGTTTTAGTTCCACCACGATTATATTTGAAATTT
ACGTGAGTGTGAGTGAGACTTGCATAAGAAAATAAAATCTTTAGTTGGGAAAAAATT
CAATAATATAAATGGGCTTGAGAAGGAAGCGAGGGATAGGCCTTTTTCTAAAATAGG
CCCATTTAAGCTATTAACAATCTTCAAAAGTACCACAGCGCTTAGGTAAAGAAAGCA
GCTGAGTTTATATATGGTTAGAGACGAAGTAGTGATTGAAACCACAATTTGCACACC
CGTTTTAGAGCTAGAAATAGCAAGTTAAAATAAGGCTAGTCCGTTATCAACTTGAAA
AAGTGGCACCGAGTCGGTGCTTTTTTTAGACCCAGCTTTCTTGTACAAAGTTGGCAT
TATCTAGAAAGCTTGAGCTC

gRNA 3 (In 10)

GAGCTCGAATTCGGATCCAGAAATCTCAAAATTCCGGCAGAACAATTTTGAATCTCG
ATCCGTAGAAACGAGACGGTCATTGTTTTAGTTCCACCACGATTATATTTGAAATTT
ACGTGAGTGTGAGTGAGACTTGCATAAGAAAATAAAATCTTTAGTTGGGAAAAAATT
CAATAATATAAATGGGCTTGAGAAGGAAGCGAGGGATAGGCCTTTTTCTAAAATAGG
CCCATTTAAGCTATTAACAATCTTCAAAAGTACCACAGCGCTTAGGTAAAGAAAGCA
GCTGAGTTTATATATGGTTAGAGACGAAGTAGTGATTGACGTGTTCATTGTTTAACC
AGTTTTAGAGCTAGAAATAGCAAGTTAAAATAAGGCTAGTCCGTTATCAACTTGAAA
AAGTGGCACCGAGTCGGTGCTTTTTTTAGACCCAGCTTTCTTGTACAAAGTTGGCAT
TATCTAGAAAGCTTGAGCTC

gRNA 4 (In 4)

GAGCTCGAATTCGGATCCAGAAATCTCAAAATTCCGGCAGAACAATTTTGAATCTCG
ATCCGTAGAAACGAGACGGTCATTGTTTTAGTTCCACCACGATTATATTTGAAATTT

ACGTGAGTGTGAGTGAGACTTGCATAAGAAAATAAAATCTTTAGTTGGGAAAAAATT
CAATAATATAAATGGGCTTGAGAAGGAAGCGAGGGATAGGCCTTTTTCTAAAATAGG
CCCATTTAAGCTATTAACAATCTTCAAAAGTACCACAGCGCTTAGGTAAAGAAAGCA
GCTGAGTTTATATATGGTTAGAGACGAAGTAGTGATTGTGGCCAAACCACCAAATTC
GTTTTAGAGCTAGAAATAGCAAGTTAAAATAAGGCTAGTCCGTTATCAACTTGAAAA
AGTGGCACCGAGTCGGTGCTTTTTTTAGACCCAGCTTTCTTGTACAAAGTTGGCATT
ATCTAGAAAGCTTGAGCTC

gRNA 5 (In 5)

GAGCTCGAATTCGGATCCAGAAATCTCAAAATTCCGGCAGAACAATTTTGAATCTCG
ATCCGTAGAAACGAGACGGTCATTGTTTTAGTTCCACCACGATTATATTTGAAATTT
ACGTGAGTGTGAGTGAGACTTGCATAAGAAAATAAAATCTTTAGTTGGGAAAAAATT
CAATAATATAAATGGGCTTGAGAAGGAAGCGAGGGATAGGCCTTTTTCTAAAATAGG
CCCATTTAAGCTATTAACAATCTTCAAAAGTACCACAGCGCTTAGGTAAAGAAAGCA
GCTGAGTTTATATATGGTTAGAGACGAAGTAGTGATTGTATAAAAAACTAATATTTG
GGTTTTAGAGCTAGAAATAGCAAGTTAAAATAAGGCTAGTCCGTTATCAACTTGAAA
AAGTGGCACCGAGTCGGTGCTTTTTTTAGACCCAGCTTTCTTGTACAAAGTTGGCAT
TATCTAGAAAGCTTGAGCTC

gRNA 6 (In 3)

GAGCTCGAATTCGGATCCAGAAATCTCAAAATTCCGGCAGAACAATTTTGAATCTCG
ATCCGTAGAAACGAGACGGTCATTGTTTTAGTTCCACCACGATTATATTTGAAATTT
ACGTGAGTGTGAGTGAGACTTGCATAAGAAAATAAAATCTTTAGTTGGGAAAAAATT
CAATAATATAAATGGGCTTGAGAAGGAAGCGAGGGATAGGCCTTTTTCTAAAATAGG
CCCATTTAAGCTATTAACAATCTTCAAAAGTACCACAGCGCTTAGGTAAAGAAAGCA
GCTGAGTTTATATATGGTTAGAGACGAAGTAGTGATTGGGTATCTTTTTGTGGGTAA
GTTTTAGAGCTAGAAATAGCAAGTTAAAATAAGGCTAGTCCGTTATCAACTTGAAAA
AGTGGCACCGAGTCGGTGCTTTTTTTAGACCCAGCTTTCTTGTACAAAGTTGGCATT
ATCTAGAAAGCTTGAGCTC

Bs3 promoter

CAAATAATGATTTTATTTTGACTGATAGTGACCTGTTCGTTGCAACAAATTGATGAG
CAATGCTTTTTTATAATGCCAACTTTGTACAAAAAAGCAGGCTTCATAGTCAAGCTA
ACGAAACTTATGCAAGGGAAATATGAAATTAGTATGCAAGTAAACTCAAAGAACTAA
TCATTGAACTGAAAGATCAATATATCAAAAAAAAAAAAAAAACAATAAAACCGTTTAA
CCGATAGATTAACCATTTCTGGTTCAGTTTATGGGTTAAACCACAATTTGCACACCC
TGGTTAAACAATGAACACGTTTGCCTGACCAATTTTATCCCTTTATCTCTAACCATC
CTCACAACTTCAAGTTATCATCCCCTTTCTCTTTTCTCCTCTTGTTCTTGTCACCCG
CTAAATCTATCAAAACACAAGTAGTCCTAGTTGCACATATATTTCACCCAGCTTTCT
TGTACAAAGTTGGCATTATAAGAAAGCATTGCTTATCAATTTGTTGCAACGAACAGG
TCACTATCAGTCAAAATAAAATCATTATTTG

Primers for dCas9 amplification

dCas9-F   5' CACCATGGACAAGAAGTATTCTATCGGACTGGCCA 3'

dCas9-R   5' CTATACCTTTCTCTTCTTTTTTGGATCTACCTTTCTC
3'

Primers for dCas9 sequencing

dCas9-F   5' CACCATGGACAAGAAGTATTCTATCGGACTGGCCA 3'

dCas9-F2   5' GAGGATCTGCTGCGGAAGCAGCGCACTTTCGA 3'

dCas9-F3   5' GAGCTCGGGTCACAGATCCTTAAAGAGCA 3'

dCas9-F4   5' GGAAGTCAAGAAGGACCTTATCATCAAGCT 3'

dCas9-R1   5' TGGGGAATGCTCCCATTGTCGAAAGTGCGCT 3'

dCas9-R2   5' AGCTGGGTGTTTTCCACCGGGTGCTC 3'

dCas9-R3   5′ TCCAGTTCGAACAGGCTATACTTTGGGAGCTT 3′

Primers for gRNA amplification and sequencing

gRNA_ampl_F   5′ GAGCTCGAATTCGGATCCAGAAATCTCA 3′

gRNA_ampl_R   5′ GAGCTCAAGCTTTCTAGATAATGCCAAC 3′

Primers for PDS amplification (for RT-PCR)

qRT-NB-PDS-303F   5′ CGTTGGGAACTGAAAGTCAAGATG 3′

qRT-NB-PDS-875R   5′ ATGGCGCAGGAAGAGCTTCAG 3′

Primers for actin1 amplification

N.B.Actin1-F   5′ TGAAGATCCTCACAGAGCCTGG 3′

N.B.Actin1-R   5′ TTGTATGTGGTCTCGTGGATTC 3′

Printed by Books on Demand GmbH, Norderstedt / Germany